Stop Faking It!

Finally Understanding Science   So You Can Teach It

SOUND

NATIONAL SCIENCE TEACHERS ASSOCIATION

Arlington, Virginia

Claire Reinburg, Director
Andrew Cocke, Associate Editor
Judy Cusick, Associate Editor
Betty Smith, Associate Editor

ART AND DESIGN Linda Olliver, Director
    Brian Diskin, Illustrator
NSTA WEB Tim Weber, Webmaster
PERIODICALS PUBLISHING Shelley Carey, Director
PRINTING AND PRODUCTION Catherine Lorrain-Hale, Director
    Nguyet Tran, Assistant Production Manager
    Jack Parker, Desktop Publishing Specialist
PUBLICATIONS OPERATIONS Hank Janowsky, Manager
sciLINKS Tyson Brown, Manager
    David Anderson, Web and Development Coordinator

NATIONAL SCIENCE TEACHERS ASSOCIATION
Gerald F. Wheeler, Executive Director
David Beacom, Publisher

Sound: Stop Faking It! *Finally Understanding Science So You Can Teach It*
    NSTA Stock Number: PB169X4
08   07   06   4   3   2

Library of Congress Cataloging-in-Publication Data
Robertson, William C.
 Sound / [by William C. Robertson, Brian Diskin].
    p. cm. — (Stop faking it!)
 ISBN 0-87355-216-4
 1. Sound—Popular works. I. Diskin, Brian. II. Title.
 QC225.3.R56 2003
 534—dc22

                          2003018807

NSTA is committed to publishing quality materials that promote the best in inquiry-based science education. However, conditions of actual use may vary and the safety procedures and practices described in this book are intended to serve only as a guide. Additional precautionary measures may be required. NSTA and the author(s) do not warrant or represent that the procedure and practices in this book meet any safety code or standard or federal, state, or local regulations. NSTA and the author(s) disclaim any liability for personal injury or damage to property arising out of or relating to the use of this book including any recommendations, instructions, or materials contained therein.

Permission is granted in advance for reproduction for purpose of classroom or workshop instruction. To request permission for other uses, send specific requests to: NSTA Press, 1840 Wilson Boulevard, Arlington, Virginia 22201-3000. Website: www.nsta.org

SCiLINKS. *Featuring sciLINKS®—a new way of connecting text and the Internet. Up-to-the-minute online content, classroom ideas, and other materials are just a click away. Go to page x to learn more about this new educational resource.*

Contents

Preface

When I was back in college, there was a course titled Physics for Poets. At a school where I taught physics, the same kind of course was referred to by the students as Football Physics. The theory behind having courses like these was that poets and/or football players, or basically anyone who wasn't a science geek, needed some kind of watered-down course because most of the people taking the course were—and this was generally true—SCARED TO DEATH OF SCIENCE.

In many years of working in education, I have found that the vast majority of elementary school teachers, parents who homeschool their kids, and parents who just want to help their kids with science homework fall into this category. Lots of "education experts" tell teachers they can solve this problem by just asking the right questions and having the kids investigate science ideas on their own. These experts say you don't need to understand the science concepts. In other words, they're telling you to fake it! Well, faking it doesn't work when it comes to teaching *anything*, so why should it work with science? Like it or not, you have to understand a subject before you can help kids with it. Ever tried teaching someone a foreign language without knowing the language?

The whole point of the *Stop Faking It!* series of books is to help you understand basic science concepts and to put to rest the myth that you can't understand science because it's too hard. If you haven't tried other ways of learning science concepts, such as looking through a college textbook, or subscribing to *Scientific American* or reading the incorrect and oversimplified science in an elementary school text, please feel free to do so and then pick up this book. If you find those other methods more enjoyable, then you really are a science geek and you ought to give this book to one of us normal folks. Just a joke, okay?

Just because this book series is intended for the nonscience geek doesn't mean it's watered-down material. Everything in here is accurate, and I'll use math when it's necessary. I will stick to the basics, though. My intent is to provide a clear picture of underlying concepts, without all the detail on units, calculations, and intimidating formulas. You can find that stuff just about anywhere. Also, I'll try to keep it lighthearted. Part of the problem with those textbooks (from elementary school through college) is that most of the authors and the teachers who use them take themselves way too seriously. I can't tell you the number of times I've written a science curriculum only to have colleagues tell me it's "too flip" or "You know, Bill, I just don't think people will get this joke." Actually, I don't really care if you get the jokes either, as long as you manage to learn some science here.

Speaking of learning the science, I have one request as you go through this book. There are two sections titled *Things to do before you read the science stuff* and *The science stuff*. The request is that you actually DO all the "things to do" when I ask you to do them. Trust me, it'll make the science easier to understand, and it's not like I'll be asking you to go out and rent a superconducting particle accelerator. Things around the house should do the trick. By the way, the book isn't organized this way (activities followed by explanations followed by applications) just because it seemed a fun thing to do. This method for presenting science concepts is based on a considerable amount of research on how people learn best and is known as the *Learning Cycle*. There are actually a number of versions of the Learning Cycle but the main idea behind them all is that we understand concepts best when we can anchor them to our previous experiences. One way to accomplish this is to provide the learner with a set of experiences and then explain relevant concepts in a way that ties the concepts to those experiences. Following that explanation with applications of the concepts helps to solidify the learner's understanding. The Learning Cycle is not the only way to teach and learn science, but it is effective in addition to being consistent with recommendations from the *National Science Education Standards* (National Research Council 1996) on how to use inquiry to teach science. (Check out Chapter 3 of the *Standards* for more on this.) In helping your children or students to understand science, or anything else for that matter, you would do well to use this same technique.

The book you have in your hands, *Sound*, starts from the basic of basics and moves to a few more complicated things. I begin with what causes sound, what sound waves are, and how to accurately describe sound waves. Then I move to how sound travels around and how different substances affect the movement of sound. A few specific topics covered are the Doppler effect, harmonics and overtones, resonance, and auditorium acoustics. There's an entire chapter on how musical instruments work, and a final chapter on listening devices, from record players to CD players, to telephones, to the human ear. I spend quite a bit of time explaining how the ear physically functions, and introduce a ton of biological vocabulary words to boot! I do not address a number of sound topics that you might find in a physical science textbook, choosing instead to provide just enough of the basics so you will be able to figure out those other concepts when you encounter them. You might also notice that this book is not laid out the way these topics might be addressed in a traditional high school or college textbook. That's because this isn't a textbook. You can learn a great deal of science from this book, but it's not a traditional approach.

One more thing to keep in mind: You actually CAN understand science. It's not that hard when you take it slowly and don't try to jam too many abstract ideas down your throat. Jamming things down your throat, by the way, seemed to be the philosophy behind just about every science course I ever took. Here's hoping this series doesn't continue that tradition.

Acknowledgments

The *Stop Faking It!* series of books is produced by the NSTA Press: Claire Reinburg, director; Andrew Cocke, project editor; Linda Olliver, art director; Catherine Lorrain-Hale, production director. Linda Olliver designed the cover from an illustration provided by artist Brian Diskin, who also created the inside illustrations.

This book was reviewed by Denise Miller (North Middle School, Illinois); Thomas Rossing (University of Edinburgh, Scotland); Tom Shoberg (Pittsburgh State University, Kansas); Daryl Taylor (Williamstown High School, New Jersey); and Kenneth Thompson (Emporia State University, Kansas).

About the Author

Bill Robertson is a science education writer, teaches online math and physics for the University of Phoenix, and reviews and edits science materials. His numerous publications cover issues ranging from conceptual understanding in physics to how to bring constructivism into the classroom. Bill has developed K–12 science curricula, teacher materials, and award-winning science kits for Biological Sciences Curriculum Study, The Wild Goose Company, and River Deep. Bill has a master's degree in physics and a Ph.D. in science education.

About the Illustrator

The soon to be out-of-debt humorous illustrator Brian Diskin grew up outside of Chicago. He graduated from Northern Illinois University with a degree in commercial illustration, after which he taught himself cartooning. His art has appeared in many books, including *The Golfer's Personal Trainer* and *5 Lines: Limericks on Ice*. You can also find his art in newspapers, on greeting cards, on T-shirts, and on refrigerators. At any given time he can be found teaching watercolors and cartooning, and hopefully working on his ever-expanding series of *Stop Faking It!* books. You can view his work at *www.briandiskin.com*.

How can you avoid searching hundreds of science web sites to locate the best sources of information on a given topic? SciLinks, created and maintained by the National Science Teachers Association (NSTA), has the answer.

In a SciLinked text, such as this one, you'll find a logo and keyword near a concept, a URL (*www.scilinks.org*), and a keyword code. Simply go to the SciLinks web site, type in the code, and receive an annotated listing of as many as 15 web pages—all of which have gone through an extensive review process conducted by a team of science educators. SciLinks is your best source of pertinent, trustworthy Internet links on subjects from astronomy to zoology.

Need more information? Take a tour—*www.scilinks.org/tour/*

Stop Children, What's That Sound?

Aging hippies might recognize the title of this chapter from an old Buffalo Springfield song. Others will just have to ignore it, and realize that this is the first chapter in a book on sound. As you go through this book, you'll be exposed to scientific models of what sound is and how it behaves. To get us started, though, we're just going to focus on what produces sound and what causes different kinds and qualities of sound—in other words, we're starting with the basics.

1

1 Chapter

Topic: What is sound?

Go to: *www.scilinks.org*

Code: SFS01

Things to do before you read the science stuff

Just so you get the idea that doing science is fun, we're going to start with a scavenger hunt. Woo hoo. Well, not a regular scavenger hunt, but a "sound" scavenger hunt. Find a bunch of things that produce sounds. Pretty easy task, because there are lots of things that produce sounds—radios, televisions, musical instruments, wind blowing through the trees, kids banging on objects, adults banging on objects, adults yelling at kids for banging on objects, and pages of this book turning.

Now follow that easy scavenger hunt with an answer to the following question. What do all of these sound-producing things have in common? In other words, is there one thing these sound producers are doing that results in sound? That's not such an easy question, so I'll help you figure it out.

Find a guitar, a violin, a piano, or some other stringed instrument. If you don't have anything like that, use a stretched rubber band. Pluck one of the strings and describe the motion of the string. Yes, it is moving pretty fast, but do your best.

Now take the cover off a stereo speaker and watch what the various little speakers inside do as music plays. See any similarity between this motion and the motion of the instrument strings that were producing sound?[1] That's about the easiest question you'll be asked in this book!

Talk while placing your fingers on the front of your neck. Does your neck feel any different when you're talking and when you're silent?

One more easy thing to do. Tap your finger lightly on a table. Produces a sound, yes? Now tap lightly on a pillow. No sound. Why the difference?

The science stuff

Vibrations produce sound. Oh my gosh, stop the presses! Yes, you probably already knew that. The vibrations that produce sound are fairly obvious when watching a plucked string or watching stereo speakers move in and out. The vibrations might not be so obvious in the case of, say, rustling leaves, but if you

---

[1] By the way, the motion of these speakers is caused by the interaction of magnets in the speakers with electrical current that comes from the amplifier. I'll discuss that a bit more in Chapter 7, but for a detailed discussion of electromagnetic interactions, look for the *Stop Faking It!* book on Electricity and Magnetism (in press).

National Science Teachers Association

watch leaves carefully as wind blows through a tree, you'll see the tiny vibra-tions. The vibrations also are not so obvious when someone speaks, but you can certainly feel the vibrations (of your vocal chords) when you place your fingers against your neck.

What about tapping your finger on a hard surface? It makes a sound, but where are the vibrations? Actually, the surface itself is vibrating. It's just that the vibrations are so small that you can't see them. You can, however, feel them. Have someone tap a table while you place your fingers lightly on the surface. You should be able to feel the vibrations. If you have a laser around, you actually *can* see the vibrations of the surface. If you were to shine a laser beam on a mirror that's on the surface, as in Figure 1.1, the reflection of the beam would dance up and down when you tapped the surface, indicating that the surface itself was vibrating.

Figure 1.1

laser beam

The spot where the laser beam hits the wall will dance up and down

mirror

Because a pillow is soft and simply yields to your finger when you tap it, there are no vibrations and hence, little or no sound.

More things to do before you read more science stuff

Go back to that stereo speaker with the cover off. Gradually change the volume of the music you might be playing, and notice any difference in the motion of the speakers. At low volumes you might not even notice any motion, but you should definitely notice the motion at high volumes.

Figure 1.2

Str. Knucklebreaker

Grab a plastic ruler and hold it on a table so that half the ruler hangs out over the edge of the table (Figure 1.2). Pluck the free end of the ruler lightly and then with a big whack. The ruler vibrates in each case, producing a sound. Notice any difference in the sound produced when you pluck it lightly and when you really hit it hard?

If you have a guitar or other stringed instrument around, pluck one of the strings lightly and then hard. You should get the same kind of difference in sound as with the ruler. And in case you haven't noticed, we're easing into the concepts slowly!

More science stuff

When you pluck a ruler softly and then hard, you are changing the size of the vibration produced in the ruler. The difference in the size of vibration is pretty obvious with the ruler and with the speakers, and maybe not so obvious with the stringed instrument. In scientific terms, we say you are changing the **amplitude** of the vibration. Larger amplitudes produce louder sounds. Notice also that changing the amplitude of vibration changes *only* the loudness, and not any other property of the sound such as pitch or quality. I'll define these terms later in the chapter.

Even more things to do before you read even more science stuff

Topic: properties of sound

Go to: *www.scilinks.org*

Code: SFS02

Back when I was a kid, one way to be really cool was to tape a playing card to your bike so the spokes on the wheel hit the card as you rode along. Being cool was easier in those days! To re-create the magic of youth, get a bike and turn it upside down so the wheels turn freely when you crank the pedals. Tape or clothespin a playing card or index card to the frame so the wheel spokes hit the card as the wheel spins.

Crank the pedals by hand and the vibrating card should make some noise. Crank the pedals at different speeds and notice what happens to the sound produced by the card. As you change the pedal speed, what do you think is happening to the speed at which the card vibrates?

Ruler-plucking time again. Place the ruler as before, so it hangs over the edge of a table. Pluck the ruler and notice the sound. Now reposition the ruler so either more or less of the ruler hangs over the edge. Pluck it again and notice any change in the sound. Practice a bit and you can play *Mary Had a Little Lamb* or *In-A-Gadda-Da-Vida* just by changing the position of the ruler as you pluck it.

Figure 1.3

baseball card
in spoke

Even more science stuff

With the card and the ruler, you produced different sounds by changing the speed at which the object vibrated. That difference in speed of vibration should have been easy to see with the ruler. A more precise way of describing the speed of vibration is to refer to the **frequency** of vibration. Frequency is defined as the number of vibrations per second. The more vibrations per second, the higher the frequency. Low-frequency vibrations produce low notes, and high-frequency vibrations produce high notes.

The perceived "lowness" or "highness" of a note is called the **pitch** of the note, so we can just state that changing the frequency of vibration changes the pitch of the sound produced.

And that's about it for the first chapter—short and sweet. It's always good to start with a confidence builder!

SC*I*LINKS.
THE WORLD'S A CLICK AWAY

Topic: sound quality

Go to: *www.scilinks.org*

Code: SFS03

Chapter Summary

Vibrating objects produce sound.

The distance through which a vibrating object moves is known as the *amplitude* of vibration.[2] The larger the amplitude of vibration, the louder the sound produced.

The number of vibrations per second is known as the *frequency* of vibration of an object. Increasing the frequency of vibration of an object increases the perceived pitch of the sound produced. Decreasing the frequency of vibration of an object decreases the perceived pitch of the sound produced.

Applications

1. One of the more annoying sounds in the world is when chalk squeaks as you write with it on a chalkboard. That sound alone is enough to make you appreciate the invention of dry-erase boards. What causes that sound? In other words, what's vibrating? Chalk squeaks when you don't slant it enough as you write, keeping it in a position that's almost straight out from the board. When you do this, the end of the chalk "catches" on the chalkboard and then slips. This causes the chalk to vibrate, giving you that high-pitched squeak. In other words, it's all in your technique!

2. While we're on annoying sounds, what about that high-pitched squeal that a mosquito makes as it does a fly-by on your ear as you're trying to get to sleep? Is that a mosquito scream or what? Nah, it's just the vibration of the mosquito's wings. Because a mosquito is tiny, with tiny wings, those wings vibrate really fast (a high frequency), producing a high-pitched sound. Contrast that sound with the sounds produced by larger insects, such as flies and bees. The larger wings on these beasts are harder to move (at least for the insects), so the vibration is slower, leading to a lower frequency and lower pitch. A discerning ear can detect the difference between large flies and small flies, and between bumblebees and honeybees.

3. If you live where it snows, you might have noticed that sometimes snow squeaks when you walk on it, and sometimes not. You might also have noticed that whether or not the snow squeaks depends on the temperature. Cold snow (around 10 degrees Fahrenheit or colder) squeaks, while warmer

---

[2] The proper definition of amplitude is actually *half* the total distance the object moves, but we can safely ignore that for the scope of what we're covering in this book.

snow doesn't. The reason is that cold snow is a bit more rigid than warm snow. When you walk on cold snow, the snow under your foot rubs against the snow around it (much like two rocks rubbing against each other). This produces vibrations, which are the squeaks. Warmer snow, being less rigid, "gives" more easily, and no vibrations are produced.

Waving Strings

Unless you really haven't been paying attention to the world around you, you know that plucked strings produce sounds. Carlos Santana demonstrates this on a regular basis. Because stringed instruments are so commonly used to produce sounds, we're going to spend a chapter figuring out what exactly those strings are doing when you pluck them. We're going to be analyzing what happens when you produce waves on strings. The waves I'll be dealing with *produce* sounds, but they are not themselves sound waves—something you have undoubtedly heard of. I'll address sound waves later, and how they relate to the waves produced on strings.

STANTON
STRINGS
ORCHESTRA

Figure 2.1

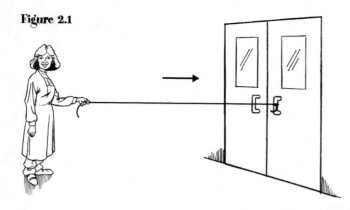

Figure 2.2

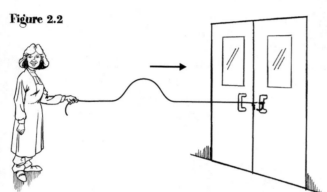

Things to do before you read the science stuff

It might seem logical that, because we're going to be talking about waves on a string, I would ask you to find a long section of string. String will work for what you're going to do in this section, but you'll need something different later in this chapter, so you might as well get that something different right now. Head to a medical supply place and buy a section of latex tubing that's at least three or four meters long.

Tie one end of the tubing to a doorknob or some other fixed object (Figure 2.1).

Quickly move the free end up and down once. You should see a wave pulse move down the tubing, as shown in Figure 2.2.

Now move the tubing up and down several times in rapid succession to see if you can get several pulses on the tubing at once (Figure 2.3).

This is a little difficult to see because you have reflected pulses heading back from the other end. Even though those reflected pulses cause trouble, try to see a difference when you make several pulses quickly, and when you make them even more quickly. Look for any difference in the distance between individual pulses.

Figure 2.3

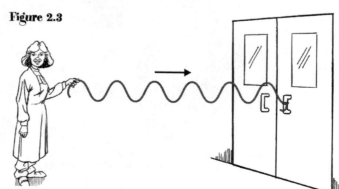

The science stuff[1]

Let's use your observations to get specific about how waves behave. First, you might have noticed that an up-and-down motion of the tubing produced waves that traveled to the side.

Figure 2.4

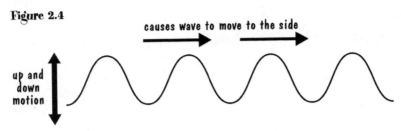

To be more precise, the waves traveled in a direction *perpendicular to* the direction of motion of whatever caused them. Waves like this are known as **transverse waves** (Figure 2.4).

The second thing you might have noticed is that the faster you move up and down to create the waves, the closer together they are (Figure 2.5). If you didn't notice that, go back and do it again.

Figure 2.5

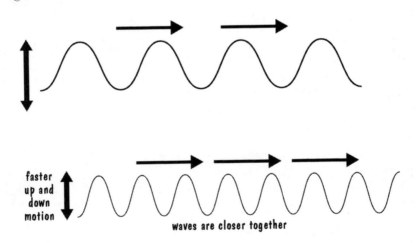

---

[1] If you have read the *Stop Faking It!* book on Light, you might recognize this section as being a whole lot like the first part of Chapter 2 in the light book. While the sections are similar, the one you're reading now goes into a bit more depth on the relationship between frequency and wavelength, and that extra depth is important for understanding how strings produce different kinds of sounds.

With smaller waves that are closer together, more of them pass a given point in a given time than do larger and farther-apart waves. In order to keep track of the size of waves, and how fast they pass a given point, we have a couple of definitions.

wavelength: The distance in which a wave repeats itself.[2] A couple of examples are shown in Figure 2.6.

frequency: The number of wavelengths that pass a given point per second.[3]

Figure 2.6

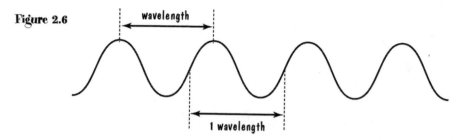

Let's assume you have a bunch of waves that all travel at the same speed, and suppose these waves are all going past you.[4]

If someone comes by and *shortens* the wavelength of these waves, what happens to the frequency?

Figure 2.7

You watch as waves go by

Figure 2.8

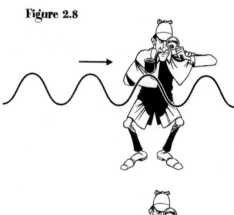

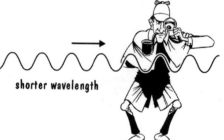

shorter wavelength

---

[2] It might be useful to think of one wavelength as the distance from one "crest" (high point on the wave) to another.

---

[3] This might look like a different definition of frequenc you in Chapter 1. There, I defined frequency as a nu second. As a wavelength passes by, though, it goes through one up-and-down motion—in other words, one vibration. So the definitions are actually the same.

---

[4] You probably didn't, or couldn't, notice, but the waves you produced on the rope in the previous section actually did travel at the same speed, regardless of how fast you moved the end of the rope. You'll just have to trust me on that one.

Well, with shorter wavelengths, more wavelengths pass you in a given time, so that means the frequency *increases*. By the same token, increasing the wavelength lowers the frequency. That gives us the following rule:

For waves traveling at the same speed, when wavelength increases, frequency decreases. When wavelength decreases, frequency increases.

Now I'd like to make the relationship between frequency and wavelength more precise. To do that, I have to define a simple thing called *velocity*. For the purposes of this book, you can think of velocity as being the same thing as *speed*.[5] The velocity of something is the amount of distance it covers in a given time. For example, you measure the velocity of a car in miles per hour. The car's velocity in miles per hour tells you how far the car will travel in one hour, assuming the car's velocity remains the same for the whole hour.

Topic: vibrations and waves

Go to: *www.scilinks.org*

Code: SFS04

Okay, let's suppose you're watching a train go by.[6] Let's say you see two cars go by every second (this is a really fast train!). Can we use that information to figure out the velocity of the train? Sure, as long as we know the length of each car. Let's suppose each car is 20 meters long. Then if two cars per second pass by us, that means 40 meters worth of train go by every second (ignoring the space between the cars).

Figure 2.9

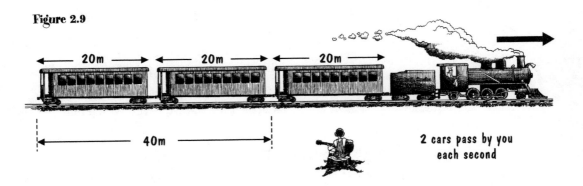

20m 20m 20m

40m

2 cars pass by you each second

---

[5] Physicists draw a clear distinction between the terms *velocity* and *speed*. For an explanation of that distinction, see the *Stop Faking It!* book on Force and Motion.

---

[6] You might be wondering why I'm talking about trains when we're supposed to be dealing with waves on a section of latex tubing. Bear with me. Once you understand what I'm about to tell you about trains, the waves on the tubing will be easier to understand.

Therefore, the train is moving with a velocity of 40 meters per second. Now I'm going to write that calculation as a relationship between the length of a train car and the number of cars that pass per second. In other words, I'm going to write a (gasp) formula that represents what we just did.

(number of cars that pass per second)(length of each car) = (velocity of the train)

Note for the math phobic

Just so you don't get totally intimidated by the symbols in an equation, here's a brief explanation. An equals sign means that whatever is on the left side of the sign is numerically the same as what's on the right-hand side. If two things are in parentheses and next to each other, as in (number of cars that pass per second)(length of each car), that means you multiply the two together. If there's a slash between those parentheses, as in (distance)/(time), you divide the first by the second. If there are just letters and no parentheses, two letters next to each other, as in vt, should be multiplied and two letters with a slash in between, as in d/t, means divide the first by the second.

Just to complete this example, let's put our numbers in.

(number of cars that pass per second)(length of each car) = (velocity of the train)

(2 cars per second)(20 meters per car) = 40 meters per second

Okay, let's apply this reasoning to waves on a string going past you. Each wave is like a train car, so we can say that

(number of waves that pass per second)(length of each wave) = (speed of the waves)

If you check back to our definition of frequency, that first term on the left is, in fact, the frequency of the waves. The length of each wave is the wavelength, so we can write

(frequency)(wavelength) = velocity of waves

I'm going to use the symbol f to represent the frequency, and the Greek letter lambda, or λ, to represent the wavelength. Representing the velocity with v, we end up with

$$f\lambda = v$$

Remember that when two letters are next to each other in an equation, you're supposed to multiply them together.

Okay, so the frequency times the wavelength equals the velocity? What's that got to do with anything? Well, first of all, this relationship holds true for all waves—the waves you produce on a string (or tubing) and also the sound waves we'll deal with in later chapters. The relationship will be important for understanding how you tune a guitar and how you get that great high-pitched voice when you inhale helium.

More things to do before you read more science stuff

Head back to that latex tubing you have tied at one end. Pull the free end until the tubing is just barely taut. It will sag just slightly. Pluck the tubing next to where you're holding it. You should get a single pulse that travels to the tied end, reflects, and heads back toward you. In fact, the pulse will reflect several times before dying out.

Figure 2.10

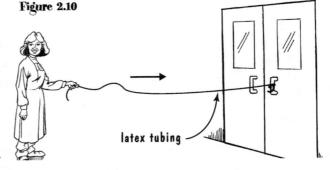

Pluck the tubing several times, and get a feel for how fast the pulse is moving. The easiest way to estimate the velocity of the pulse is to take note of how often the reflected pulses hit your end of the tubing. Now pull the tubing tighter and pluck it again. Again note the velocity of the pulse. Keep pulling the tubing tighter and tighter, checking the velocity of the plucked pulse each time. Notice a pattern? Does how tight you pull the tubing affect the velocity of the pulse?

latex tubing

More science stuff

To answer that previous question, yes, the tightness of the tubing affects the velocity of the pulse. The scientific term for the tightness of a piece of tubing or a piece of string (strings behave just like your tubing) is called **tension**. So what we have learned is that increasing the tension in a string increases the velocity of pulses that travel along the string. Because waves on a string are just extended pulses, this also holds true for waves on a string. The higher the tension, the greater the velocity of the waves.

A more difficult thing to demonstrate (which is why you won't do that) is the fact that the heavier a string is, the *slower* waves travel along the string. That sort of makes sense, because it will be more difficult for wave motion to move a heavy string than a light string. And just to be more precise, we can say that the *density* of a string (how much material there is in a given length of string) affects the speed at which waves travel along the string.

Before moving on, let's recap. We now know that there is a definite relationship between frequency, wavelength, and velocity ($f\lambda = v$); and that the velocity can be affected by changing either the tension or the density of the string. All this info will come together in the last section of this chapter.

Even more things to do before you read even more science stuff

Because I know you don't want to gather any more materials for this chapter, we're going to keep using that same section of latex tubing tied at one end.

Figure 2.11

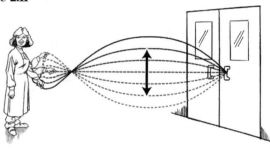

Doing your best to keep the tension in the tubing constant, move the free end up and down, slowly at first, and then faster. With a little practice, you can produce the patterns shown in Figure 2.11. These are called **standing waves**, and are actually the net result when waves going down the tubing add to the waves reflected from the other end of the tubing. The reason they're called standing waves is certain parts of the tubing actually remain still, and if the vibrations are really fast (high frequencies), the parts that are moving even seem to be standing still.[7]

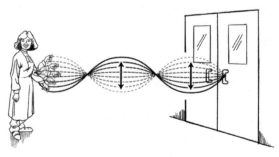

Take note of the frequencies (how fast you move the tubing up and down) at which these patterns occur.[8]

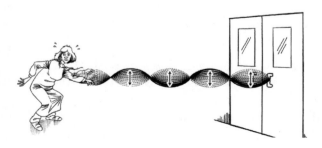

[7] You won't be able to achieve such high frequencies with just your hand, but feel free to try!

[8] Of course, you won't be able to determine the exact frequency. Just get a feel for how fast you move the tubing for each pattern, so you'll be able to compare with later frequencies.

Now increase the tension in the tubing (pull it tighter) while keeping the length of the tubing you're using the same. Because the tubing stretches, this means you have to "choke up" on your end to keep the length the same. See Figure 2.12. Create the patterns in Figure 2.11 again, and notice whether or not the frequencies at which the patterns occur change.

You have just determined what happens to the frequencies of the patterns when you keep the length the same and change the tension. Now I want you to keep the tension the same and change the length. In other words, create the standing wave patterns using about half the tubing and then with the entire length of the tubing, doing your best to keep the tension constant, as shown in Figure 2.13. Again, notice any difference in the frequencies at which the standing wave patterns occur.

Okay, I lied. You do need to get something besides the tubing. Find a guitar, violin, or other stringed instrument. Pluck one of the strings and notice the pitch (frequency) of the sound

Topic: waves

Go to: *www.scilinks.org*

Code: SFS05

Figure 2.12

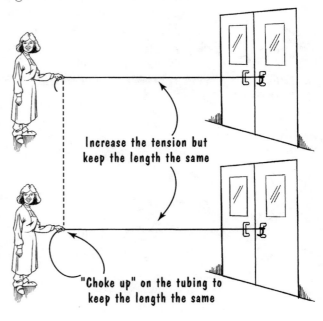

Increase the tension but keep the length the same

"Choke up" on the tubing to keep the length the same

Figure 2.13

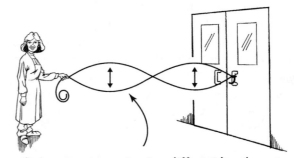

Same pattern, same tension, different lengths

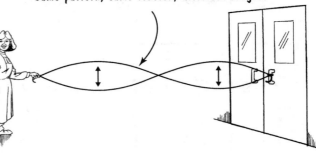

Figure 2.14

Entire
string vibrates

Shortened
string vibrates

produced. Now tighten (increase the tension in) the string you plucked (you use those little knobs on the end of the instrument to do this!) and pluck it again. Notice any change in pitch? Use that same knob to lower the tension (loosen the string) and see what happens to the pitch when you pluck the string again.

Next you're going to see (or rather, hear) what happens when you change the length of the string being plucked. To change the length, you simply push down with your finger on one of the *frets* of the instrument (see Figure 2.14) and pluck the resulting shortened segment of the string.

Mess around with different lengths of each string (hold the strings down at different frets) to figure out the relationship between the length of the vibrating string and the pitch, or frequency, of the sound produced.

Even more science stuff

Before explaining what you just observed, I'm going to analyze standing wave patterns in more detail. When you move the free end of the tubing up and down, you are sending waves toward the tied end of the tubing. Those waves reflect off the tied end, and what you end up with on the tubing is a combination of those original and reflected waves. For most of the frequencies, this combination of original and reflected waves results in a jumbled mess. For certain frequencies, though, you get those cool standing wave patterns. It's as though the tubing "likes" those certain frequencies, and responds strongly to them. The frequencies at which a string or section of tubing "likes" to vibrate are known as **resonant frequencies**. When you pluck a string, it tends to vibrate at those resonant frequencies (standing wave patterns), and since plucked strings produce music, it's worth studying those resonant frequencies and their corresponding standing wave patterns. Let's look at what one of those standing waves looks like in a series of snapshots (Figure 2.15).

Figure 2.15

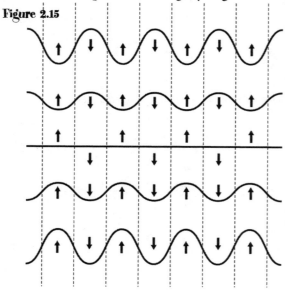

Figure 2.16

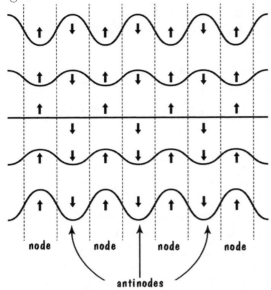

node   node   node   node

antinodes

Notice that some parts of the tubing move up and down a whole bunch, but some spots don't move at all. The parts that move the most are called **antinodes** and the parts that don't move at all are called **nodes**, as labeled in Figure 2.16.

Also notice that the length of the tubing that's vibrating determines the wavelength of the standing waves in a particular pattern. For example, the pattern shown in Figure 2.17 has a wavelength that's equal to half the length of the tubing. As long as you don't change the length, there's no way to change the wavelength of that standing wave pattern.

Armed with that information, we can now understand what happened when you kept the length of the tubing the same but changed the tension in the tubing. Because you kept the length the same, the wavelength of

each standing wave pattern remained constant even though you changed the tension (Figure 2.18).

We learned earlier, however, that increasing the tension in the tubing does change something; it changes the *velocity* at which waves move along the tubing. We also learned earlier that there is a special relationship between the frequency, wavelength, and velocity of waves traveling along tubing or string, namely $f\lambda = v$.

Figure 2.17

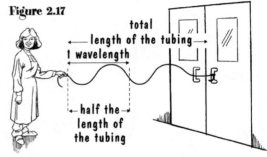

total length of the tubing

1 wavelength

half the length of the tubing

Figure 2.18

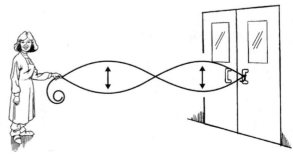

Tension increases, but, the wavelength stays the same

This relationship is an *equality,* meaning that the left side must always have the same value as the right side. It helps to think of this equality as a teeter-totter,[9] with *f*λ on one side and *v* on the other. In order to maintain the equality, the teeter-totter must always be in balance, as in Figure 2.19.

Figure 2.19

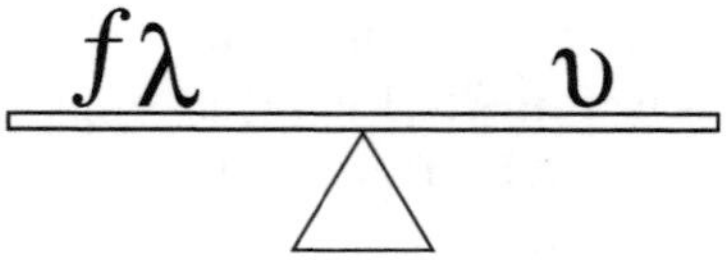

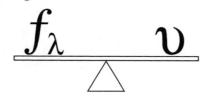

Figure 2.20

Figure 2.21

If we increase *v* (by increasing the tension), that means the value of the right side increases. We can represent the increase in *v* by making that letter larger on the right side of the teeter-totter. This unbalances the teeter-totter, meaning the two sides are no longer equal, shown in Figure 2.20.

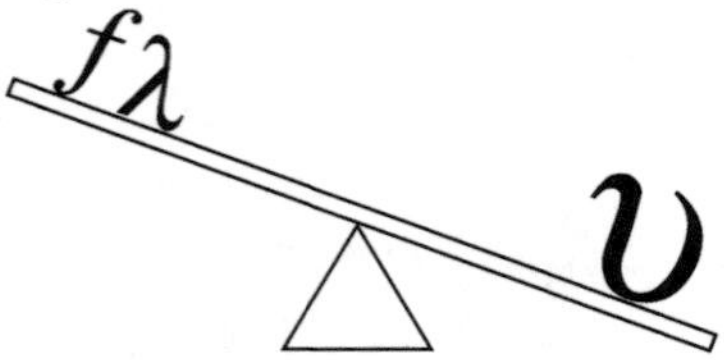

Okay, so the left side must also increase in value for the teeter-totter to remain in balance. However, we know that because we're keeping the length of the tubing constant, the wavelength of any particular standing wave pattern must remain constant. If the wavelength (λ) remains constant, the only way for the left side to increase in value is for the frequency (*f*) to increase in value, as shown in Figure 2.21.

To sum up, increasing the tension in the tubing increases the velocity at which waves travel along the tubing, which increases the frequency at which a particular standing wave pattern will occur. And of course, you observed that directly in the previous section. To put our result more simply:

Increasing the tension in a string or tubing increases the frequency at which standing waves occur on the string or tubing. By the same token, decreasing the tension in a string or tubing decreases the frequency at which standing waves occur on the string or tubing.

The above relationship is really important for tuning stringed instruments. To make a string vibrate at higher frequencies when you pluck it, you increase the tension in the string. To make a string vibrate at lower frequencies when you pluck it, you decrease the tension in the string. And that's why you fiddle with those knobs on the end of a guitar in order to tune it.

Now let's move on to what happens when you change the length of a vibrating string or tubing, while keeping the tension in the string or tubing constant. If

---

[9] I used this teeter-totter analogy to explain Newton's second law in the *Stop Faking It!* book on Force and Motion. If you read and understood that explanation, then this one should be a piece of cake!

you keep the tension constant, that means the velocity of waves moving along the string remains constant. Therefore, the right-hand side of the equation $f\lambda = v$ remains constant. What happens to the other two terms, f and λ? To see what happens to the wavelength of a particular standing wave pattern when you change the length, take a look at Figure 2.22. This figure shows that increasing the length of the vibrating string or tubing increases the wavelength of a particular standing wave pattern, and de-

Figure 2.22

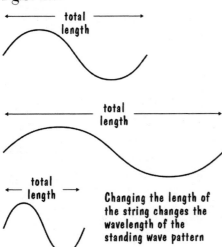

total length

total length

total length

Changing the length of the string changes the wavelength of the standing wave pattern

creasing the length of the vibrating string decreases the wavelength of a particular standing wave pattern.

Because the right hand side of $f\lambda = v$ remains constant (because the velocity remains constant), the left-hand side must also remain constant in our new (different length) situation. But one of the terms on the left, the wavelength, changes when we change the

Figure 2.23

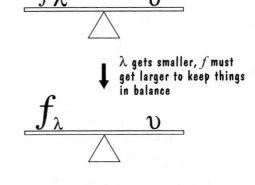

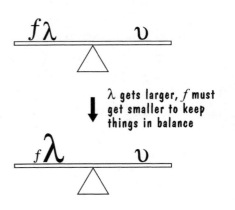

λ gets smaller, f must get larger to keep things in balance

λ gets larger, f must get smaller to keep things in balance

length. What does that mean about the other term on the left, the frequency? Well, it also must change. To be more specific, if the wavelength increases, the frequency must decrease. If the wavelength decreases, the frequency must increase. Figure 2.23 illustrates this with the teeter-totter.

So the result is:

Increasing the length of a string or tubing decreases the frequency at which standing waves occur on the string or tubing. By the same token, decreasing the length of a string or tubing increases the frequency at which standing waves occur on the string or tubing.

This helps us understand what happens when you push down on the fret of a guitar, making the length of the string that's vibrating shorter. Shortening the length makes λ smaller. Because v remains the same (you haven't changed the tension), f must get larger. Higher f means a higher frequency, and a correspondingly higher pitch.

Okay, enough about waves on strings. In the next chapter, we'll deal with sound waves, and in chapter 4, we'll discover the connection between waves on strings and sound waves. I know, you can hardly wait.

Chapter Summary

Waves on a string, which move perpendicular to the direction of the motion that caused the waves, are known as **transverse waves**.

The **wavelength** of a wave is the distance in which the wave repeats itself—the distance from one crest to another.

The **frequency** of a wave is the number of waves that pass a given point per second.

There is a special relationship between the frequency of waves on a string, their wavelength, and the velocity with which they move along the string. That relationship is $f\lambda = v$, where f is the frequency, λ is the wavelength, and v is the velocity.

Changing the tension (tightness) of a string changes the velocity at which waves move along the string. Increased tension means increased velocity and decreased tension means decreased velocity.

A string of a given length, tension, and density responds strongly when you apply waves of certain frequencies. These frequencies are known as **resonant frequencies**, and the wave patterns these frequencies produce on the string are called **standing waves**.

When you pluck a string, it tends to vibrate at its resonant frequencies.

Because of the relationship between tension and velocity of waves on a string, and because of the relationship $f\lambda = v$, you can change the frequencies at which a plucked string vibrates by changing the *tension* in the string.

Because the length of a string determines the wavelengths of standing waves that are produced on a plucked string, and because of the relationship $f\lambda = v$, you can change the frequencies at which a plucked string vibrates by changing the *length* of the string.

Applications

1. I told you that the density of a string affects how fast waves travel along the string. Let's see that in action. Find a guitar or a piano (not an electric piano). If the piano is your musical choice, raise the lid so you can see the strings of the piano. Notice that the strings that produce low notes (meaning low pitch and low frequency) are a whole bunch thicker than the strings that produce the high notes (Figure 2.24). Same thing holds for a guitar, but the difference in thickness isn't quite as pronounced.

   A given section of a thick string weighs more than that same length of thin string. This means that the thicker a string is, the more *dense* it is.[10] So the strings that produce low notes are denser than the strings that produce high notes. That should mean that waves travel slower along the low-note strings than along the high-note strings. We won't be able to see that directly, but we can *infer* it using the following argument.

   $f\lambda = v$ works for all waves on all strings, so let's apply it to two strings that are the same length but have different densities. Because the strings are the same length, the wavelengths of any standing waves produced on the different strings will have the same wavelength (Figure 2.25).

   The string that produces the low note is denser than the string that produces the high note. Therefore, waves travel *slower* on the

Figure 2.24

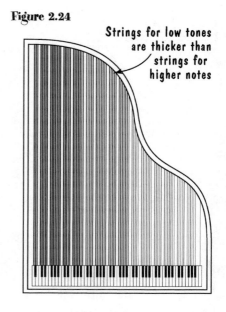

Strings for low tones are thicker than strings for higher notes

Figure 2.25

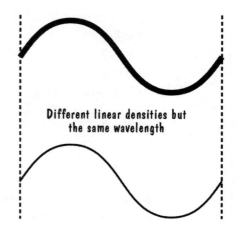

Different linear densities but the same wavelength

---

[10] I'm assuming here that the two strings in question are made of the same material. Also, I'm talking about *linear* density here—how much mass the string has *per unit length*. If the strings are made of the same material, the *volume* density (what one usually thinks of when talking about density) of the strings actually is the same.

Figure 2.26

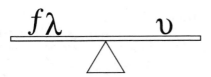

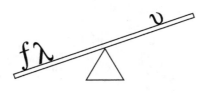

↓ υ gets smaller

low-note string than on the high-note string. That means v is a smaller number for this string. (See Figure 2.26.)

From $f\lambda = v$ then, we know that f is lower for the low-note string in order to keep our teeter-totter in balance (Figure 2.27).

Figure 2.27

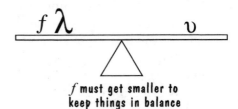

f must get smaller to
keep things in balance

And this means all is right with the world, because lower frequencies mean lower notes. A heavier string will produce notes that are lower than the same length string that is lighter. Of course, this assumes that each string has the same tension. If they have different tensions, that will affect the velocity with which waves move along the strings and mess up our analysis.

2. Place a rubber band between two fingers and pluck it. Notice the pitch of the sound produced. Now stretch the rubber band and pluck it again. Any change in pitch? You can investigate how the pitch changes for all sorts of different stretches of the rubber band if you can find a friend to do the plucking (Figure 2.28).

How does the amount you stretch the rubber band affect the pitch? Does this make sense? Cue the Jeopardy theme while you figure out your answer. Da da da da, da da da. . . . Da da da da *da*, dadadadada. . . . Okay, I'm back. What happened is that the pitch of a plucked rubber band doesn't change very much, if at all, when you stretch it. Why, you ask? Because many different things are changing when you stretch a rubber band. First, you are increasing the tension.

From our earlier examples, we know that this should increase the pitch. How-

Figure 2.28

DUDLEY
the
Wonder
Dog

ever, you are also changing the length of the rubber band, making it longer. Longer plucked strings tend to vibrate at lower frequencies, meaning a lower pitch. So the increase in tension tends to increase the frequency and pitch, and the increase in length tends to decrease the frequency and pitch. To mess things up even more, you are making the rubber band thinner as you stretch it. This lowers the density and tends to make the pitch higher. All in all, these different effects tend to cancel one another out, and you end up with very little change in pitch.

How Sound Gets Around

S o far we've been messing around with waves on strings and tubing. Sound travels in waves, but those waves aren't exactly like waves on a string. One nice thing, though, is that sound waves have all the same relationships between frequency, wavelength, and velocity that waves on a string have. What that means is what we've covered so far isn't wasted. Kind of reassuring, huh? At any rate, maybe it's time to find out what sound waves are like.

Things to do before you read the science stuff

The first thing I want you to do requires special equipment—a bell jar and a vacuum pump. If you have access to that equipment (try a friend who teaches high school or college physics), great. If not, just follow along as I explain what's to be done and what will happen when you do it.

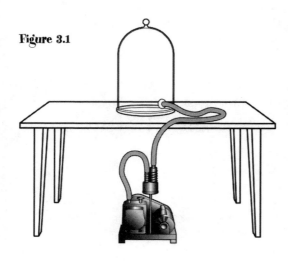

Figure 3.1

Figure 3.1 shows a bell jar attached to a vacuum pump.

When you fire up the pump, it removes air from inside the bell jar. The longer the pump runs, the less air there is inside the bell jar. Of course, the pump can't remove all the air from the bell jar (which would be a perfect vacuum), but it can remove quite a bit.[1]

What you do is get a ringing alarm clock or a buzzer or any other continuous maker of noise and place it inside the bell jar. The bell jar muffles some of the sound, but you can still hear the ringing. Start up the vacuum pump and listen. The longer the pump runs, the quieter the ringing is. Eventually, you won't hear the alarm clock at all. What really makes an impression is when you have an alarm clock with the two bells on top and a clapper that vibrates in between them. After the pump has been going a while, you hear nothing, but the clapper is still hitting the bells! Turn the pump off and let air into the bell jar, and you can hear the alarm clock again.[2]

Grab a friend and find yourself the side of a building, as in Figure 3.2. Stand maybe 15 meters away from your friend and have your friend tap the side of the building loud enough that you can hear it. Might help if your friend uses a stick

---

[1] The normal atmospheric pressure makes it basically impossible to create a perfect vacuum (no air molecules at all in the bell jar). There's a great Far Side cartoon that has a woman pushing a vacuum cleaner along a path in the forest. The caption reads: "The woods were dark and foreboding, and Alice sensed that sinister eyes were watching her every step. Worst of all, she knew that Nature abhorred a vacuum."

---

[2] As long as you have access to a bell jar and vacuum pump, you simply *must* place a few marshmallows inside the jar and start up the pump. What happens then, and after you let air back into the bell jar, is just too cool.

or other hard object to do the tapping. Now place one ear against the side of the building and plug the other one. Have your friend tap again. Do you still hear the tapping? Is it louder or softer than when you were just standing by the building?

If it's summertime, put on your swimsuit and head with your friend to the pool. If it's winter and you don't have access to an indoor pool, you can do this next thing by filling the bathtub with

Figure 3.2

"If I break a nail..."

Figure 3.3

S-H-A-R-K...?!

water and sticking your head in. No, not nearly as much fun as trekking to the pool! Let's assume you made it to the pool. Give your friend a stick or something else with which to hit the side of the pool. While your friend is hitting the side of the pool, you dunk your head underwater (Figure 3.3). Can you hear the banging? Does it sound the same as when your head is above water?

The science stuff

In each situation described above, there was a source of sound (the alarm clock, the person banging the side of the building or the pool) and a receiver of sound (you). Since you heard the sound in each case (except when the vacuum pump did its thing), it makes sense that the sound *traveled* to you. What you found out is that sound can travel through air, through wood or aluminum siding, and through water. What sound *can't* travel through is *nothing*. When the vacuum pump removed a lot of the air from inside the bell jar, you couldn't hear a thing. So unlike light, which can travel through empty space,[3] sound must travel through something. Scientists say that sound requires a **medium** through which to travel. Here the word medium has nothing to do with middle, average, or psychics, but rather refers to some kind of substance, such as air, wood, or water. In the next section,

---

[3] See the *Stop Faking It!* book on Light.

we'll explore exactly how sound travels through a medium. Once you understand that mechanism, you'll understand why sound requires a medium.

More things to do before you read more science stuff

Buy a Slinky or steal one from your kids. Lay it out on a smooth surface such as a linoleum floor. By jiggling one end of the Slinky back and forth, create *transverse*[4] waves that travel down the Slinky (See Figure 3.4).

Now, instead of moving the Slinky perpendicular to the direction of travel of the waves, push and pull the Slinky, as in Figure 3.5. As you do this, watch the rest of

Figure 3.4

I really need to clean under that fridge....

Figure 3.5

the Slinky for any motion that you might call "wavelike." This will be easier to observe if you start with just one pulse, and then work up to several pulses.

Repeat this kind of motion a number of times, watching the Slinky carefully. What exactly is the Slinky doing as a wave pulse passes by?

More science stuff

Hopefully you noticed that when you moved the end of the Slinky back and forth along the length of the Slinky, you produced a wave that consisted of the Slinky being compressed and then stretched as the wave went along. Maybe you

---

[4] Remember that transverse waves are like the ones you create on a string—the wave moves perpendicular to the motion of your hand.

saw something similar to Figure 3.6? If not, go back and mess around with the Slinky until you get that kind of motion.

Figure 3.6

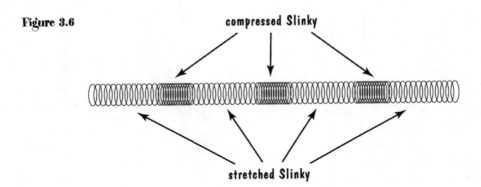

compressed Slinky

stretched Slinky

Waves like this, in which the motion of the material itself is in the same direction as the direction of travel of the waves, are called **longitudinal waves**.[5]

Figure 3.7

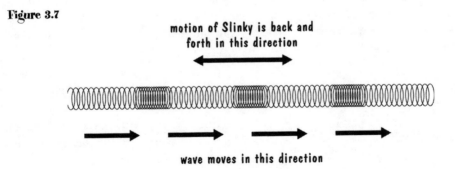

motion of Slinky is back and
forth in this direction

wave moves in this direction

Now why do you suppose I'd introduce longitudinal waves? Could it be that sound waves are longitudinal waves? Um, yeah. Let's first look at sound waves that travel through air, and let's suppose that these waves are caused by a vibrating string. Now air is made up of very tiny little things called molecules, which are moving all over the place.[6] Each time the string moves forward, it compresses

---

[5] Compare this to transverse waves on a Slinky, in which the motion of the Slinky is *perpendicular* to the direction of travel of the waves, as shown in Figure 3.4.

---

[6] An important question you should ask at this point is, "How do we know that air is made up of tiny molecules?" You could answer, "Because I learned that in third grade," but there are better reasons. I don't have the space to go into that now, but rest assured I'll answer that question in another *Stop Faking It!* book.

the air molecules in front of it (see Figure 3.8). Those molecules push the molecules in front of them, which push the molecules in front of them, which push the molecules . . . you get the picture. What you end up with is a series of wave pulses, each pulse generated by the forward movement of the string.

Because the vibration of the string is a regular back and forth motion, all those wave pulses taken together make up complete waves, which we conveniently call *sound waves*. Because the string motion is three dimensional, the sound waves it produces don't move forward just in one direction, but in all directions from the string. That's why you hear the vibrating string no matter where you're standing in relation to it. Of course, some sound producers create sound waves predominantly in one direction. Stereo speakers are an example of this, where the sound is much louder in front of the speaker than in back of the speaker.[7] Check out Figure 3.9.

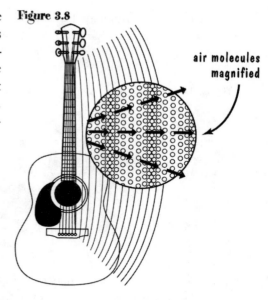

Figure 3.8

air molecules magnified

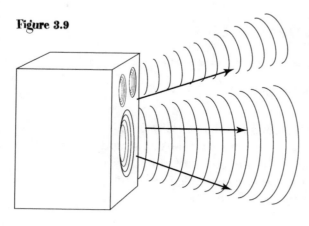

Figure 3.9

Most of the sound moves out from the front of the speaker

This next point might be obvious, but just in case it isn't, I'll explain it. When a string vibrates at a particular frequency, it produces sound waves that have that *same frequency*. Makes sense, because it's the vibrations of the string that create the sound waves. And now maybe you can see why I spent so much time covering waves on strings. There's this very close connection between the vibrating string that produces sound, and the sound waves themselves—they have the same frequency.

Sound waves travel through liquids and solids in the same way they travel through air. The only difference is that

---

[7] Hence the popularity of using a large number of speakers in "surround sound" systems.

the molecules in liquids and solids are closer together and have stronger bonds[8] than the molecules in air. It's still a situation of molecules pushing other molecules, which push other molecules, and so on, as the sound waves move through the medium.

Even more things to do before you read even more science stuff

Get a friend and a hammer (or other metal object) and find a very long metal railing or a chain-link fence. A school yard might be a good place to find such a thing. Get as far away from your friend as possible (you should be at least 50 meters away), with the fence or railing covering the distance between the two of you. What I did when figuring out the activity was use the top railing of a chain-link fence and made sure there were no gates between my friend (actually my daughter!) and me.

Figure 3.10

Your friend is hitting the top rail of the fence with a hammer

You are 50 yards away listening

Have your friend whack the railing with the hammer. You should see the hammer hit before you hear the sound. Have your friend do this a bunch of times while you take note of the fact that there's a time interval there between you seeing the hammer hit the railing and you hearing the sound.

Now put one ear up to the railing or the fence and plug your other ear. Have your friend hit the railing with the metal object again, making sure you can see the hitting motion even though you have your ear to the railing. Because sound travels through metal, you should hear, with your ear that's pressed against the railing, the sound of the metal object hitting the railing. Is there a delay between the hitting and hearing this time?

---

[8] In fact, the molecules in air are essentially independent without any bonds between them.

Your friend should hit the railing several times so you can get a feel for how long this time delay is, if there is a delay at all. Compare this time delay with the one you got when you just listened, instead of having your ear up to the railing.

Up for a little more fence whacking? This time, hold one ear up against the railing, but leave the other ear unplugged. Have your friend hit the fence with the metal object again. Listen carefully as he or she does this several times. You should hear *two* sounds, one right after the other. What do you suppose is causing those two sounds you hear?

Leave the poor fence alone and position yourself anywhere from 30 to 100 meters from a building. If you happen to be in a school yard for your fence abuse, that school building near you should do just fine. Shout at the building and listen for the echo. What do you suppose causes the time delay between your shout and the sound of the echo?

Head back inside and find your Slinky. Hold one end of the Slinky so it touches your ear and let the other end dangle free. Hit the free end with a pencil or any other hard object. If you don't hear some really cool sounds, adjust the position of the Slinky on your ear and try again. It should sound a bit like Luke and Darth Vader going at each other.

Even more science stuff

I'll start by letting you know that light travels really, really, really fast. It travels so fast that, when you're 50 meters away and watching as someone hits a fence railing with a hammer, you see the hit at almost exactly the same time as the hit actually happens. There is a slight delay, but it's too short for you to notice. You should have noticed, however, a definite delay between the hit and your hearing of the sound. That's because sound waves travel through air a whole lot slower than light travels through air. At room temperature, on a dry day, sound waves travel through air at about 344 meters per second. Now that's not exactly slow (more than three football fields in one second), but it's not nearly as fast as light travels (300,000,000 meters per second, or about seven times around the Earth in one second!). So, sound waves in air travel much slower than light travels in air, which is why there was a delay between seeing the hit and hearing the hit.

SCI LINKS.
THE WORLD'S A CLICK AWAY

Topic: speed of sound

Go to: *www.scilinks.org*

Code: SFS07

How about when you put your ear up to the railing? By plugging your other ear, you ensured that the sound you heard from the hit came through the railing and not the air. If you noticed a delay between the hit and the sound, then you have good observational skills. Most people would say that the hit and the sound happened at about the same time. Why a very short, or possibly no delay? Because sound travels through metal *much* faster than it travels through air. In fact, the speed of sound through steel is about 5,000 meters per second. Compare that with 344 meters per second for sound traveling through air.

Now on to what you heard with one ear to the railing and the other ear unplugged. You should have heard two sounds. One came from the sound of the hit traveling through the railing (you hear this first because sound travels through the railing much faster than it travels through air) and the other came from the sound reaching you by traveling through the air. All of this leads us to the following conclusion:

The speed at which sound waves travel depends on the medium in which the waves travel. If you change the medium, or change the properties of the medium, you change the speed at which sound travels.

That might not seem like the most startling revelation in the world, but it will help us understand many practical things having to do with sound, including the production of music and being able to sound like a Munchkin when you inhale helium. In the meantime, you might wonder why exactly sound travels faster in some mediums than in others. The situation is a bit complicated, but it has a lot to do with how much a substance "gives" when you try to push or pull it. Scientists call this the *compressibility* of the substance. If the substance "gives" a lot, you might guess that the substance would be sluggish in responding to sound waves. If the substance is more rigid, you might expect that the substance would transport sound waves more efficiently. It's not unlike a string with low tension transmitting waves slower than a string with high tension.

Let's move on to echoes. Basically, an **echo** demonstrates that sound waves reflect off things, and also demonstrates again that it takes sound waves a bit of time to travel from one place to another (there's a delay between your shout and the echo). If you've spent any kind of time producing echoes, you might also know that some materials reflect sound better than others. A brick building reflects just fine, but a mountain doesn't reflect so well unless there's a sheer rock cliff on that mountain. It turns out that some materials (rock, brick, metal) reflect sound waves quite well, while other materials (dirt and trees, curtains, foam rubber) do not reflect sound waves well. When sound waves don't reflect, we say they are *absorbed* by the surface.

Finally, what caused those nifty sounds when you hit the Slinky while one end was next to your ear? Echoes, of course. When you hit one end of the Slinky, sound waves travel from one end to the other, reflecting back and forth. Because the sound waves travel much faster in the Slinky than they do in air, the echoes come fast and furious, leading to sounds that mimic light sabers.

And even more things to do before you read even more science stuff

We've discussed how sound gets around, and how it travels at different speeds in different mediums. With all this talk about sound waves moving, we haven't addressed what happens when the source of the sound is moving. Time to do that.

Prepare to have the neighbors question your sanity. For this activity, you need a car, two assistants, and a whistle. One of the assistants needs to be old enough to drive. Take your equipment and assistants and find an empty stretch of road that's maybe an eighth of a mile long. A regular neighborhood street will work. You're going to stand at the side of the road while the two assistants drive by at a speed of at least 20 miles per hour. One of the assistants drives while the other leans out the window, blowing the whistle with a long, steady, blow. The whistle needs to be blowing before the car reaches you, as the car passes, and after the car goes by. That means the whistle-blower needs a lot of wind.[9]

Now that you know the setup, go ahead and do it. Take note of the pitch of the whistle both before and after the car passes you. The pitch of the whistle should be higher as the car approaches and lower as the car moves away from you. If you didn't notice the difference, try again, making sure the person blowing the whistle does so with a long, steady tone. Extra runs also give the neighbors something more to talk about.

And even more science stuff

What you just heard is known as the **Doppler effect**, named after an Austrian physicist named—surprise—Christian Doppler (the "Christian effect" might have been too confusing a reference). To see what causes the Doppler effect, first consider a stationary source of sound, such as the whistle. Sound waves travel out in all directions from the whistle. These sound waves are shown as "wave fronts" heading outwards in Figure 3.11.

Figure 3.11

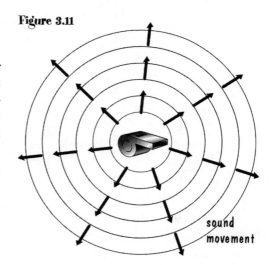

sound movement

---

[9] If you only have one assistant, he or she could just lean on the horn as the car passes. This might, however, be an unpopular option with the neighbors.

The speed at which these wave fronts travel is determined by the medium in which they travel, in this case, air.

Now suppose the whistle is moving as it blows. It still sends out wave fronts, and these wave fronts travel at the same speed they did when the whistle was stationary. Remember, the speed of the sound waves depends on the *medium* in which they travel, not on the speed of the source of sound. Once a sound wave is emitted, it travels at the speed dictated by the medium. In other words, you can't make someone hear you any faster if you move towards him or her as you shout!

The motion of the sound source *does* affect the sound, though. When the source is moving, it catches up to the sound waves in front of it and runs away from the sound waves behind it. This situation is shown in Figure 3.12. Note the difference between this and Figure 3.11.

The motion of the source (the whistle) has caused the sound waves in front of it to bunch up and the sound waves behind it to spread apart. The bunching up of sound waves means more waves pass a given point in a given amount of time than would if the waves weren't bunched up. That means someone in *front* of a moving sound source will hear a *higher frequency* than if the source were stationary. Also, someone behind a moving source will hear a *lower frequency* than if the source were stationary. Higher frequency means higher pitch and lower frequency means lower pitch. And that's exactly what the Doppler effect entails. The Doppler effect is really easy to hear at a racetrack, what with the high speeds of the cars and the loud whines of the engines.

SCI
LINKS.
THE WORLD'S A CLICK AWAY

Topic: Doppler effect

Go to: *www.scilinks.org*

Code: SFS08

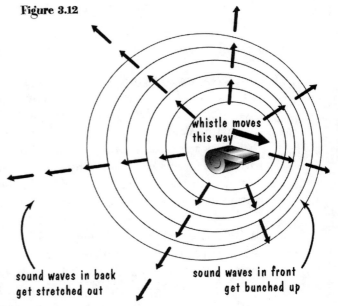

Figure 3.12

whistle moves this way

sound waves in back get stretched out

sound waves in front get bunched up

Chapter Summary

Sound requires a medium—such as air, water, or metal—in which to travel. Sound cannot travel through a vacuum (empty space).

Sound waves are longitudinal waves, in which the direction of motion of the medium is the same as the overall direction of motion of the waves.

Sound waves tend to travel outward in all directions from the source of the sound.

Sound waves travel at different speeds in different mediums.

Sound waves reflect from some surfaces and don't reflect (are absorbed) when they hit other surfaces.

When a sound source moves toward you, you hear a higher frequency than that produced by the sound source. When a sound source moves away from you, you hear a lower frequency than that produced by the sound source. This is known as the Doppler effect.

Applications

1. There's a way to figure out how far away a lightning strike is from you. What you do is count the number of seconds between seeing the lightning and hearing the thunder. Divide that count by five, and the result tells you how many miles away the lighting strike was. So, for example, if you count to five before hearing the thunder, the lightning that caused the thunder is 5 ÷ 5, or 1 mile away. Okay fine. Why does this method work? Actually, it's pretty simple. For starters, we assume that the light from the lightning gets to you instantaneously. That's not a bad assumption because light travels so fast. Then all we have to do is figure out how long it takes sound to travel a mile. A mile is equal to 1,600 meters, and sound travels about 344 meters per second. Traveling at 344 meters per second, sound covers 688 meters (that's twice 344) in two seconds, 1,032 meters in three seconds, 1,376 meters in four seconds, and 1,720 meters in five seconds. If we fudge just a bit, and realize that most people can't count out seconds exactly, this means that it takes sound about five seconds to travel one mile (1,600 meters).[10] So, if it takes sound five seconds to travel a mile, for every count of five between seeing lightning and hearing the thunder, that adds a mile to the distance the lighting is away from you. If you get to a count of two before hearing the

---

[10] In case that just confused you, all I'm saying is that 1,720 meters is pretty darned close to 1,600 meters, which is a mile.

thunder, the lightning strike is 2/5 of a mile away from you. Time to put down the golf clubs and head for cover!

2. If you want to mix culinary arts and science, this is your chance. Get a long, rectangular pan, such as you use for baking bread loaves, and use it for a Jell-O mold. Fill it with Jell-O (the usual recipe—hot and cold water, no bananas or cottage cheese)—and when it's set, plop it out onto a tray. Whack the Jell-O at one end and you'll see some great longitudinal waves travel along the Jell-O. Okay, Jell-O might not qualify as culinary arts, but it's still a cool thing to watch.

3. What movie dealing with submarines is complete without that "boop . . . boop . . . boop . . . " sound of the **sonar** detecting an enemy submarine or an iceberg? To figure out how sonar works, imagine throwing a ball against a wall while you're blindfolded. If you throw the ball at about the same speed each time, you can tell how far away you are from the wall by noting how long it takes for the ball to hit the wall. Sonar is pretty much the same process. We know how fast sound travels in seawater. We send out a sound pulse, and measure how long it takes for those sound waves to leave us, hit the enemy submarine, and bounce off and come back to us. Knowing the time it took and knowing the speed of sound in water, we can then figure out how far away the object is. Great, but we can get even more information from the echo we hear. If we're not moving, and the echo has a higher frequency than the sound we sent out, we know that the enemy sub is moving towards us. This info comes courtesy of the Doppler effect. I won't go into the details, but you can figure out what echo frequency you should hear when your sub is moving and when the enemy sub is moving. In other words, you not only know where that enemy sub is; you also know how fast and in what direction it's moving. And in case you're wondering, there's a Doppler effect for light waves, and that's how the State Patrol uses radar to figure out just how much you owe for ignoring those little signs that tell you how fast you can go in your car. By the way, *sonar* stands for SOund NAvigation Ranging.

4. Ever wonder how oil companies know where to drill for oil? They use something very similar to sonar. They set off a charge of dynamite. The sound produced travels into the Earth below and reflects off all kinds of layers of rock (see Figure 3.13). A set of sound-detecting instruments called *geophones*, which are sitting on the surface, pick up and record all the reflections coming back. A computer analyzes the data from the geophones and produces a picture of the rock formations below the surface.

Figure 3.13

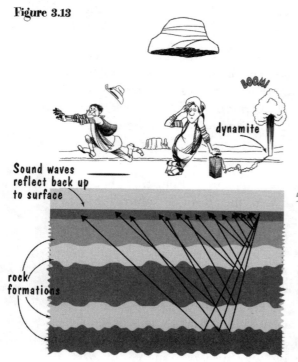

dynamite

Sound waves reflect back up to surface

rock formations

Certain kinds of rock formations indicate the presence of oil or gas deposits, and that means a likely place to drill. This bit of information comes to you courtesy of a summer I spent planting geophones in the ground while hiking through swamps, fighting mosquitoes, and being thankful I never got picked to be on the dynamite crew.

5. There's a part in the Jules Verne book *Journey to the Center of the Earth* where one group of people can hear another group talking, even though they're miles away from each other. That's actually possible given just the right setup. If your surroundings are structured just right, even whispers can travel a great distance, reflecting off walls and traveling through tunnels by reflection.

6. One place you might have heard the word Doppler is on your local weather report. Weather people often refer to the pictures provided by "Doppler radar." Well, that radar uses the Doppler effect to determine how fast, and in what direction, cloud formations are moving.

7. Isn't it cool when a planet explodes in *Star Wars* and you hear a giant explosion? Yep, but totally bogus. Because sound waves need a medium in which to travel, they can't travel through empty space. When things blow up in space, here's what you hear: ....

Harmonic Convergence

I chose the title of this chapter based on the fact that this chapter deals with a sound phenomenon known as *harmonics*. Of course, harmonic convergence actually refers to a day back in 1987 when there was a particular planetary alignment. New Agers went crazy over the event, and if you want to get a taste of what it was like, enter harmonic convergence into your favorite Internet search engine and sit back and enjoy. My wife and I celebrated the day. I played my harmonica and my wife used a converging lens to focus the Sun's rays and burn a piece of paper in the backyard.

Enough silliness. This chapter adds to our previous discussion of resonance by expanding our list of vibrating things to metal rods and columns of air.

Things to do before you read the science stuff

Head to the hardware store and buy two metal rods (if the rods are threaded, that's okay) that have the same diameter but different length. You'll likely find lengths of 1 and 3 feet and diameters of 1/4 or 3/8 inch.[1] So if you get two 1/4-inch thick rods, one of length 1 foot and one of length 3 feet, you'll be set. Make sure the rods are made of the same material. If you don't have any string at home, get yourself a ball of that while you're out.

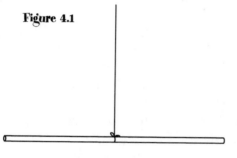

Figure 4.1

When you're back home, tie a piece of string around the center of one of the metal rods, so the rod can hang freely when you hold the string. See Figure 4.1.[2]

While holding the suspended rod, hit it with the other rod. You should get a nice, clear tone. Kinda pretty, huh? Try this with each of the rods, noting how length affects the pitch you hear. Right about now you might be realizing that a banged metal rod is a whole lot like a plucked string. I'll be using the same kind of analysis I used for plucked strings to analyze these rods in the next section. If you want extra credit, though, go ahead and see if you can use $f\lambda = v$ to explain how the length and thickness of the rods affected pitch.

Get three or four empty pop or beer bottles (glass, not plastic) and add different amounts of water to each bottle. The amounts should vary from about 1/4 full to 3/4 full. Blow across the top of each bottle (see Figure 4.2), so you get a distinct pitch from each bottle.

Notice any relationship between the amount of water in the bottle and the pitch you hear? We're looking for a general pattern here and not necessarily an exact mathematical relationship.

Figure 4.2

---

[1] In this book series, we usually try to stick with metric units. In this case, however, the metal rods you find at the hardware store will have measurements in English units. You can find a better time to work on conversion of units than when you're out shopping.

---

[2] If you're using a threaded metal bar, it's pretty easy to get the bar to balance horizontally. A smooth bar might be more difficult. If you have trouble, try using a bent paper clip at the end of the string on which to balance the bar.

Just to mess up what you think you know, take a spoon and *hit* the bottles instead of blowing over the tops (Figure 4.3). Now what's the relationship between the amount of water and the pitch you hear?

Figure 4.3

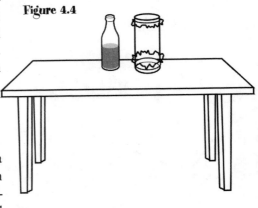

Craft time. Get an empty toilet paper tube, two balloons, a couple of rubber bands, and a small bit of fine-grained sand. Cut the balloons open so you can stretch them across the ends of the toilet paper tube. Use the rubber bands to keep the balloon pieces in place, stretching the pieces as tight as you can, as shown in Figure 4.4.

Take a pencil and tap lightly on one end of the fancy drum you just made. Quality sound it ain't, but what do you expect given the expense of the materials? Sprinkle some sand on top of the drum and tap continuously until the sand arranges itself into a definite pattern. Try tapping on different parts of the balloon piece and see if the pattern changes.

Figure 4.4

The science stuff

Let's deal with the metal rods first. Hitting a metal rod is a lot like plucking a string. When you hit the rod, it vibrates at its resonant frequencies,[3] and that vibration produces sound waves. Just as with a plucked string, the frequencies at which a metal rod vibrates depend on its length. Also just as with a plucked string, we can figure out, using $f\lambda = v$, why the different lengths produce the frequencies they do.

Also, just as with a string, changing only the length does not change the velocity at which waves move along the rod. Changing the length *does*, however, change the wavelength of any standing waves that are set up on the rod. This is illustrated in Figure 4.5. You might notice that this standing wave looks different from the ones in Chapter 2. In particular, the ends of the rod vibrate up and down (they're antinodes) and the center remains still (it's a node). The reason for the difference is that by hanging the rod at its center, you leave the ends free

---

[3] If you forget what a resonant frequency is, head back to Chapter 2 for a quick review.

Figure 4.5

Increasing the length of the rod increases
the wavelength of the standing wave

Figure 4.6

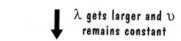

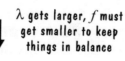

λ gets larger and v
remains constant

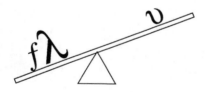

λ gets larger, f must
get smaller to keep
things in balance

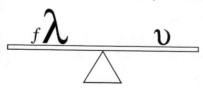

to vibrate, whereas with a vibrating string, the ends of the string are fixed and not free to vibrate.[4]

So changing the length of the rod changes the wavelength of the standing waves set up on the rod. An increase in the length of the rod means an increase in the wavelength. Because the velocity stays the same as you change the length, the right-hand side of the equation $f\lambda = v$ remains constant. With the wavelength of the waves going up (because the rod is longer), the frequency must go down to keep things in balance. Lower frequency means lower pitch, so the longer rod produces a lower note. Here are those teeter-totters again as a reminder of how we get to that lower frequency. Notice that you only have an equality when the teeter-totter is balanced. The second teeter-totter represents a situation where the two sides are *not* equal (Figure 4.6.)

Now on to blowing over the tops of bottles. Other than helping you train for a career in a jug band, what good is all that? Well, it takes us from vibrating strings and rods to vibrating columns of air. When you blow over the top of a bottle, you set up standing waves inside the bottle. Those standing waves, however, are *longitudinal* waves, and it's the air inside the bottle that's vibrating.

To get an idea of what longitudinal standing waves are like, get that Slinky you used in Chapter 2 and find a smooth sur-

[4] You might also notice that this drawing is greatly exaggerated. The rod doesn't vibrate with this large an amplitude. If we drew the vibration to scale, though, you wouldn't be able to see the standing wave!

face. Hold one end of the Slinky still and use your other hand to create longitudinal waves at the other end (see Figure 4.7). Slowly increase the frequency of your motion until you get the following pattern in the Slinky.

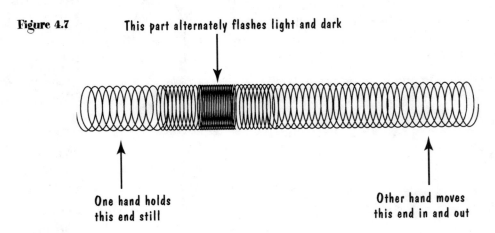

Figure 4.7   This part alternately flashes light and dark

One hand holds
this end still

Other hand moves
this end in and out

What you've created with this pattern is a longitudinal standing wave. The end you are holding still is a *node* (it's not moving at all), and the end you are moving back and forth is an *antinode* (it's moving a lot). The part of the Slinky that keeps flashing as a dark region is also an *antinode*. There, the Slinky alternately compresses fully and expands fully. It becomes dark when fully compressed. Now because longitudinal standing waves are difficult to draw, we can represent it with a drawing of a transverse standing wave, keeping in mind that we're really talking about a longitudinal wave. Check out Figure 4.8.

So now we have longitudinal standing waves inside a bottle, and it's the air molecules inside the bottle that are vibrating back and forth. Let's see if we can figure out where the nodes and antinodes are inside the bottle. First think about the opening. You're blowing across the top of the bottle, setting up the vibrations. That means the molecules at the opening are moving back and forth quite a bit, so that must be an antinode.[5] At the water's surface (or the

Figure 4.8

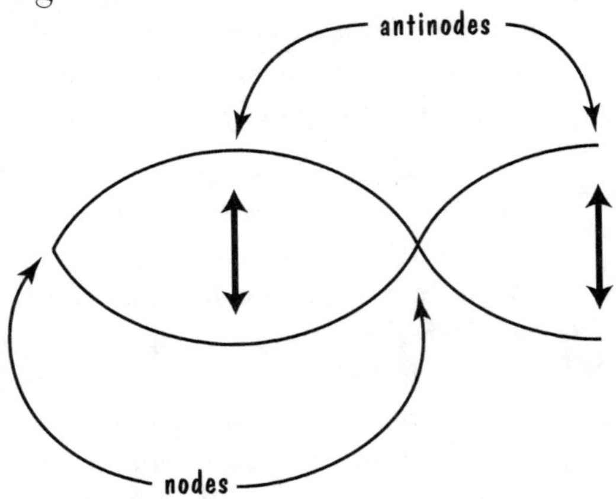

antinodes

nodes

A transverse wave representation of the
longitudinal standing wave pattern on the Slinky

------

[5] Actually, the true antinode of this standing wave isn't exactly at the opening. It's close to the opening, though, and it's not worth complicating things by going into detail.

bottom of the bottle if it's empty), the air molecules *aren't* free to move around because it's a dead end. Therefore, the water's surface has to be a node, where the molecules don't move at all. So we have the situation in Figure 4.9.

Okay, so we know that at one end we have a node and at another we have an antinode. We don't really know what might be happening in between, except that there are a lot of possibilities, as shown in

Figure 4.9

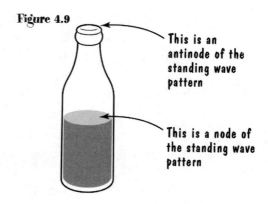

This is an antinode of the standing wave pattern

This is a node of the standing wave pattern

Figure 4.10

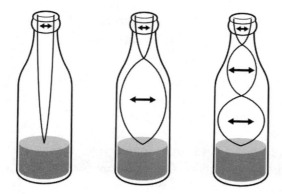

Possible standing wave patterns for longitudinal waves set up inside a bottle

Figure 4.10. Keep in mind that we're still using transverse wave patterns to represent longitudinal waves.

Maybe you recall back in Chapter 2 that when you pluck a string, it vibrates at a whole bunch of resonant frequencies. Well, it's the same when you blow across the top of a bottle. The air inside, given the space it has, tends to vibrate at all sorts of resonant frequencies. This means that all of those patterns shown in Figure 4.10 are happening at once. Of course, we only hear one of the frequencies (we only hear one note), which is an issue we'll address in the next couple of sections of this chapter. For now, maybe you can just accept that when you blow across the top of a bottle filled with water, you hear a certain resonant frequency that is determined by how much water is in the bottle.

Did you accept that? Good. Now let's figure out why you hear different notes when you have different amounts of water in the bottles. It's not really all that difficult to figure out, because longitudinal waves (the waves of air inside the bottle) behave just like transverse waves on a string. That means that our good old friend $f\lambda = v$ applies. When we change the level of water in the bottle, that doesn't change anything about the air inside the bottle. That means the *medium* in which these waves travel remains the same, and also means that the velocity of the waves remains the same. So, the right-hand side of $f\lambda = v$ stays constant when we change the level of water in the bottle. What doesn't stay constant when you change the level of water is the length in which the standing waves operate. And just as with plucked strings and banged metal rods, changing the length changes the wavelength of the standing waves.

Back to the teeter-totter. We start with f and λ on one side of the equation, with v on the other. Everything's in balance for a given length of the column of air that's vibrating. If we increase the length of the column of air, that increases λ (see Figure 4.11). That unbalances the teeter-totter, as in Figure 4.12.

To keep the teeter-totter in balance, which has to happen because that's the way waves behave, f has to get smaller, as shown in Figure 4.13.

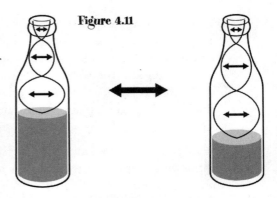

Figure 4.11

Increasing the length of the column of air increases the wavelength of the standing waves

Figure 4.12

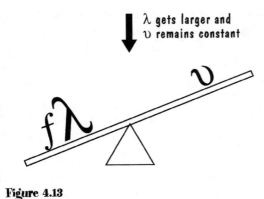

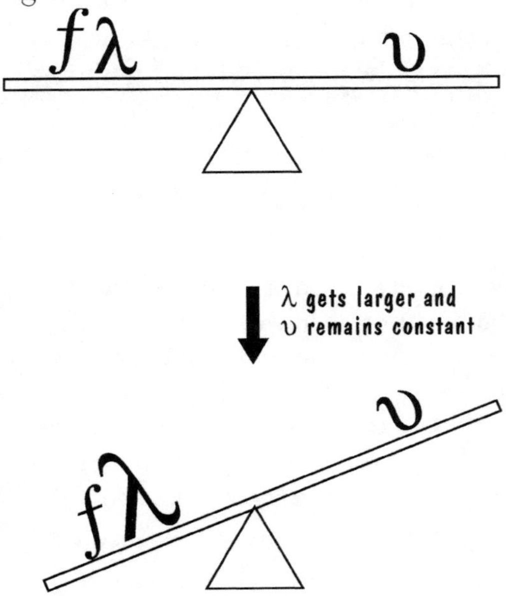

λ **gets larger and v remains constant**

Figure 4.13

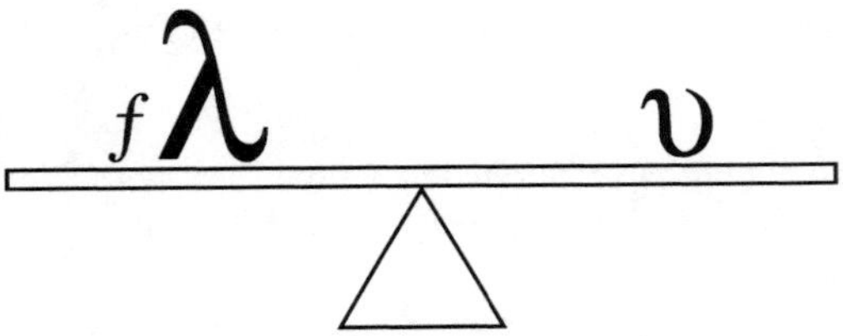

λ **gets larger, f must get smaller to keep things in balance**

What all that means is, as the water level in the bottle gets lower (the column of vibrating air gets longer, as does the wavelength of the standing wave), so does the frequency. Less water means a lower pitch. Likewise, more water means a higher pitch.

At this point, it might seem like we're beating this topic to death. We've gone over how length changes pitch a bunch of times, and you might be thinking "Okay, I get it already!" Well, if you get it, good. It turns out that the process of translating physical situations like this into math, and then back to a physical conclusion, is a tough thing for many people to follow. So, the more times we do it, the better. Also, once you have a solid grasp of this concept, you'll understand the chapter on musical instruments without breaking a sweat. And of course, that's what this whole book series is all about—helping you understand basic concepts so thoroughly that teaching those concepts to others just won't be a problem.

When you banged the bottles with a spoon, you got opposite results, right? The less water in the bottle, the higher the pitch. There's a good reason for you getting opposite results. When you hit the bottle instead of blowing on it, the bottle itself and the water inside (not the air inside) are what vibrate. The more water in the bottle, the slower the whole thing tends to vibrate. It's a lot like thicker strings on a piano or guitar vibrating at lower frequencies than thinner strings.

Finally, let's look at that makeshift drum. The pattern that the sand made when you tapped on the balloon segment is called a **Chladni figure**.[6] The pattern you saw was probably irregular, but if you take a real drumhead and repeat the activity, you'll get symmetrical and intricate patterns. What's happening is that tapping the balloon segment sets up standing waves in that segment. The only difference between these and other standing waves we've been dealing with is that the drumhead standing waves are two dimensional (the entire drum surface moves up and down like mountains and valleys). Certain parts of the drumhead move up and down a lot (these are antinodes) and certain parts don't move at all (these are the nodes). The sand collects at the nodes, and this leads to the pattern you saw.

More things to do before you read more science stuff

Grab the metal rods you used earlier. You're going to use one rod to hit the other and make various notes. You won't need the string tied around the center of the rod. Instead, grasp the rod firmly at the center using just your thumb and forefinger and hit it with the other rod. If you're right at the center, you'll hear the same note you heard when dangling this same rod by a string. This note won't be quite as clear as when the rod was attached to the string because your fingers are keeping the rod from vibrating freely. If you're slightly off center, you'll hear a dull thud. Adjust your fingers until you hear a relatively clear note.

Now hold the rod about 1/4 of the way along its length, still using just your thumb and forefinger. Rap it again with the other rod.[7] Adjust your fingers until you hear another relatively clear note. This note should have a higher pitch than the one you heard earlier. Repeat that last step, but this time try holding the rod about 1/6 of the way from the end. You should hear an even higher note. Try

---

[6] Ernest Florens Friedrich Chladni from Saxony is often appropriately referred to as "the Father of Acoustics." Seems his demonstrations of metal plates stroked with violin bows used to be quite the hit 200 years ago.

---

[7] You can hit the rod on either side of the spot where you're holding it. Where you hit the rod shouldn't make a difference in the resultant sound.

holding the rod at all sorts of different positions, and see if you can get all sorts of different notes when you bang it.

If you have a recorder, a flute, or some other wind instrument, simply play a few different notes by covering and uncovering different holes. After doing that, see if you can figure out how holding a rod in different places and banging it, and playing a wind instrument while covering and uncovering different holes, are the same kind of process. If you can't figure it out, have no fear; I'm going to explain it in the next section!

More science stuff

I've already told you that when you pluck a string or hit a metal rod, it vibrates with characteristic resonant frequencies that correspond to standing waves. Figure 2.11 shows what those waves look like for a plucked string. The ends of the string are nodes (no movement) because the ends are tied down. For a metal rod that's free at both ends, the resonant frequencies look a bit different. Because the ends of the rod are free to move about, the ends are *antinodes* of the standing wave patterns. Therefore, the possible standing wave patterns for a metal rod look like those in Figure 4.14.

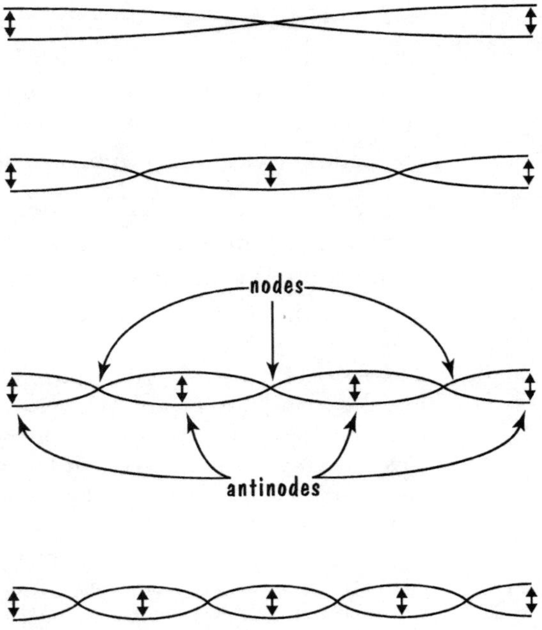

It's time to add names to these different patterns of standing waves. The lowest frequency standing wave (the first one in figure 4.14) is called the **fundamental**. The next lowest frequency standing wave (the second one in Figure 4.14) is called the *first harmonic.* The third lowest is the *second harmonic,* the fourth lowest the *third harmonic,* and so on. All of the harmonic frequencies are known collectively as **overtones**.

When you pluck a string or hit a rod, the frequency that dominates the vibration is the fundamental, so that's the note you hear. That means when you hit a metal rod that's hanging from a string, the note you hear corresponds to the first standing wave shown in Figure 4.14. What changes when you *hold* the rod in the center instead of suspend it from a string? Answer: nothing. Because the center of the rod is a

Figure 4.14

nodes

antinodes

Topic: harmonics

Go to: www.scilinks.org

Code: SFS09

node for the fundamental frequency (meaning that part doesn't vibrate), holding the rod at that point doesn't change anything. The rod is still free to vibrate with the standing wave pattern corresponding to the fundamental.

Okay, what happens when you hold the rod about a fourth of the way from the end and hit it? You hear a higher note, but why? Well, by holding the rod a fourth of the way from the end, you are *preventing* the rod from vibrating at its fundamental frequency. You're keeping that part of the rod from vibrating or, in other words, you're forcing that point on the rod to be a *node*. If you look at the second drawing in Figure 4.14, you'll see that this standing wave pattern does have a node at a point about 1/4 of the way from the end of the rod. Therefore, the rod *is* allowed to vibrate at the frequency corresponding to the first harmonic. Therefore, we hear that frequency. Notice that because the first harmonic has a shorter wavelength than the fundamental, the corresponding note has a higher frequency and thus a higher pitch. When you prevent the rod from vibrating at certain frequencies, we say you are **damping**[8] those frequencies.

So, by holding the rod at different places, you damp some frequencies and allow others to occur, leading to different notes when you hit the rod.[9]

Let's just do a quick review before moving on. When you hit a metal rod (or pluck a string), the rod vibrates with all sorts of standing wave patterns, and these patterns correspond to particular frequencies and thus particular notes. The dominant frequency you hear is the fundamental. By holding the rod at different places, you can force certain parts of the rod to be vibration nodes—places where the rod doesn't vibrate at all. This can damp out certain frequencies, leading to the rod producing different notes.

Right about now you might be thinking you never saw someone standing on the corner playing a metal rod and begging for loose change, so let's move on to a more common musical instrument such as a flute or a recorder. If you have kids in elementary school, you know what a recorder is. If your kids are as

---

[8] I have to admit I don't know the origin of the word *damping*, and my dictionary wasn't much help. I do know that the pedal you use to soften sounds on a piano is known as the damper pedal. Note, however, that the damper pedal on a piano damps (reduces in amplitude) *all* frequencies produced by the piano strings, not just some of them.

---

[9] By holding the rod at a point one fourth of the way along the rod, you are separating the rod into a long end and a short end. You might wonder what is preventing the long end of the rod from vibrating at a longer wavelength then the short end of the rod. The reason that can't happen is that, even though you are holding the rod at a particular point and forcing that point to be a node, the long and short ends still are coupled together strongly. They can't vibrate independently.

irresponsible as my daughter, you've bought several of them. For the great unwashed out there, just think of a recorder as a simple kind of flute. You blow in one end, the other end is open, and there are a bunch of holes along the length that you can cover with your fingers. It looks something like Figure 4.15.

Figure 4.15

When you blow on a recorder, you set up standing waves in the column of air inside the recorder. These are longitudinal waves, just like the waves you set up when you blow across the top of a bottle. In this case, though, there are antinodes of vibration at both ends of the recorder. The end you're blowing on is an anti-node because you're causing the air there to move a whole bunch. The other end is open, so air at that point is free to move about, making it an antinode also. Let's see . . . antinodes at both ends . . . sounds like a vibrating metal rod, no? Yep. The air inside the recorder vibrates with the same standing wave patterns as the metal rod. Of course, these are longitudinal instead of transverse waves, but just as we did with the bottles, we'll represent those longitudinal waves as if they are transverse waves. So, blowing into a recorder with all the holes covered sets up a whole bunch of standing waves like the ones in Figure 4.16. And just as with the metal rod, the note you hear corresponds to the fundamental frequency.

Figure 4.16

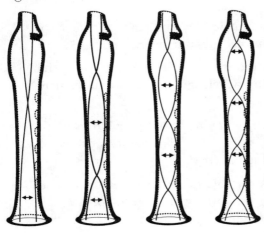

Okay, so what happens when you start uncovering holes? When you grabbed the metal rod in various places, you forced those places to be nodes in the standing wave patterns. With the recorder, by uncovering holes, you force those places to be antinodes, because opening those spots to the open air allows the air inside to vibrate freely. By forcing certain places to be antinodes, you make some standing wave patterns impossible, while allowing others. For example, uncovering the hole shown in Figure 4.17 makes the air inside the recorder vibrate with a frequency that is much higher than the fundamental frequency.

Figure 4.17

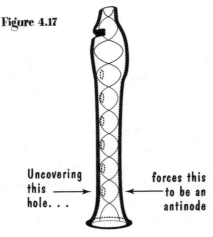

Uncovering this hole . . . → ← forces this to be an antinode

You can also *damp* certain frequencies on a drum by tapping the drum in different places or simply by placing your hand in different places. Your balloon drum is too crude for this kind of thing to work, but the next time you watch someone play drums, especially bongo drums, notice how the sound changes when the drum is hit in different places.

Chapter Summary

When you hit a metal rod, it vibrates with characteristic resonant frequencies. The resonant frequencies produced are transverse standing wave patterns.

When you cause a column of air to vibrate, it also vibrates with characteristic resonant frequencies. The resonant frequencies produced are longitudinal standing wave patterns.

When you hit a drumhead, that drumhead vibrates with characteristic resonant frequencies. These are two-dimensional standing wave patterns.

The lowest frequency (longest wavelength) standing wave is known as the *fundamental*. Higher frequencies (smaller wavelengths) than the fundamental are known as *harmonics*.

You can change the sounds produced by vibrating rods and columns of air by changing their lengths. The relationship $f\lambda = v$ helps you determine exactly how the sounds change with changing length.

You can cause a metal rod, a string, or a column of air to produce overtones—frequencies that correspond to standing wave patterns other than the fundamental pattern—by forcing certain parts of the standing wave pattern to be nodes or antinodes.

Applications

1. When you talk, your vocal chords vibrate and set up a column of vibrating air in your throat. You change the pitch of the sounds you make in many different ways. You can increase the tension in your vocal chords or change the effective length of the vibrating column of air by using more or less of your mouth or chest in producing sounds. Another way is to change the *medium* in which the sound waves travel. In other words, change from a vibrating air column to a vibrating air and helium column. Inhale a big whiff of helium from a helium-filled balloon, and all of a sudden you're talking like you should be following the yellow brick road. The reason? Sound travels faster in helium than in regular air. That means the right side of $f\lambda = v$ increases. In order for the relationship to remain true, f must also increase,

so your voice comes out at a higher frequency than normal. By the way, be careful when inhaling helium. Too much and you can suffocate from lack of oxygen in your lungs!

2. Here's a great and cheap demonstration that will baffle those around you.[10] All you need is a cup of instant hot chocolate or instant coffee. The cup should be ceramic. Right after you stir in the chocolate or coffee, begin tapping on the side of the cup with a metal spoon. Listen carefully as you tap, and you'll notice that the pitch of the sound produced gets higher and higher. Stir up the liquid and start tapping again, and the pitch will return to its original low note and then get higher again. Is this cool or what? Well sure, it's cool, but why does it happen? Tapping on the edge of the cup sets up standing waves in the cup and the liquid. Just as with all other waves, these waves obey the relationship $f\lambda = v$. Notice that λ doesn't change as you tap the cup, because the size of the cup and the amount of liquid in the cup don't change. It turns out that the velocity at which waves travel through the liquid *does* change. When you stir up the liquid, you cause extra air to dissolve into the liquid. As the cup sits, this dissolved air gradually escapes from the liquid. The less air in the liquid, the *faster* waves travel through the liquid.[11] This means that the right-hand side of $f\lambda = v$ increases. Because λ remains constant, f has to increase. An increase in f means a higher pitch. Stirring the liquid a second time dissolves more air in the liquid, starting the process over again.

3. When you pluck something, it tends to vibrate at its resonant frequencies. You can excite resonant frequencies in other ways, too. For example, find a wineglass, wet the tip of your finger, and slowly move the tip of your finger around the rim of the glass. Pretty soon the glass will start to "sing." This can be really pretty or really annoying, depending on your perspective. What's happening is that tiny vibrations between your finger and the rim of the glass are causing the glass plus water to vibrate at its fundamental frequency. By changing the amount of water in the glass, you change the fundamental frequency and therefore the pitch you hear. Every time I make glasses sing like this, I picture my friend, his mom, and me creating our own symphony with glasses, fish bowls, and jars all filled with different amounts of water. That was in sixth grade, I believe.

---

[10] I stole this idea from a book by Jearl Walker, titled *The Flying Circus of Physics*. This book is a great source of interesting physical phenomena. The explanations are brief, but after studying my book series, you should be able to understand most of them.

---

[11] This should make sense. Sound travels through liquid much faster than it does through air. Therefore, the less air in the liquid, the faster the waves travel through it.

4. You can hear the overtones on a guitar string in much the same way as you do on a metal rod. You simply place your fingers very lightly on top of the string in just the right places and pluck the string with your other hand. By placing your fingers lightly on the string, you damp the fundamental and possibly other lower frequencies, while still allowing the entire string to vibrate. Overtones have a light and delicate sound that you just can't produce by plucking the guitar strings normally.

5. In a rather morose application of sound waves, I recall that recently (this will date the book!), a tape was released that was purported to be the voice of Osama bin Laden. We were told that experts would analyze the audiotape and let us know whether or not this was, in fact, the voice of bin Laden. How could they figure that out? With voice recognition software, of course! But what the heck is that? Well, remember that I told you that anything that produces sound produces all sorts of frequencies—the fundamental, and all the harmonics. And those frequencies produced depend very much on what is producing them. In this case we're talking about the throat, lungs, and mouth of the person talking. It stands to reason that each person, with different physical characteristics, produces different sets of sound frequencies. With the help of a bit of mathematical analysis known as **Fourier analysis,** it's possible to break down a complex sound, such as someone's voice, into all the different frequencies that make up the sound. With Fourier analysis and the help of electronics that help us do the math, we can essentially establish a "voiceprint," not unlike a fingerprint, that is unique to each person.

Waves Do Basic Math—
Adding and Subtracting

S o far we've been dealing with single sources of sound, and how they produce the pitches you hear. But you don't go to the symphony to hear a single oboe, and when you hear a solo guitarist, he or she doesn't keep your attention long by playing only one note at a time. So maybe it's a good idea to find out what happens when different sound waves mix together. Okay, let's do that.

Things to do before you read the science stuff

Set up your home sound system or a boom box so two of the speakers are right next to each other and facing the same direction (you do need two speakers for this, so a regular clock radio won't work).[1] Place the speakers so you can be about 3 meters away from them with your ears at the level of the speakers. You'll have to get down on hands and knees if the speakers are on the floor (Figure 5.1).

Figure 5.1

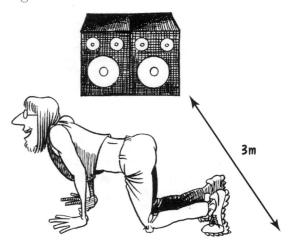

3m

Do whatever you have to in order to pick up AM radio signals. If you have a mono/stereo button on your sound system, set it for mono. Now tune the receiver so you're in between stations and picking up that really annoying weeeeooooo sort of high-pitched hum that you would normally try to avoid. This annoying hum is easier to find with a dial tuner than with a push-button digital tuner. If all you have is the latter, just do the best you can to find a constant pitch hum with as little static, talking, or music as possible.

Figure 5.2

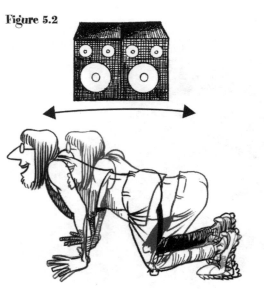

With one ear toward the speakers, move your head *slowly* back and forth in the direction shown in Figure 5.2. Concentrate on the annoying hum as you do this, and notice any change in volume. Is there also an overall pattern to the changes in volume? Say yes.

Now retune the radio so you get a much higher or much lower pitch of an annoying sound. Repeat what you already did, moving your head back

---

[1] If you have read and done the activities in the *Stop Faking It!* book on Light, you might recognize the activity that is to follow. That's because almost the same activity is in both books (the second part in this book is different)! Sound waves and light waves behave similarly.

and forth, and see whether you notice any difference in the pattern. Try all sorts of different pitches (frequencies) of annoying sounds. In case you're having trouble figuring out any pattern changes, you're trying to see how pitch affects the distance between successive loud or soft spots as you move your head back and forth.

The science stuff

I'm going to explain why you got the patterns you did with sound waves coming from two speakers. Once again, I'm going to use *transverse* waves to illustrate what's going on, even though sound waves are longitudinal waves. The reason for this approach is simple—transverse waves are easier to visualize and to draw. We'll start by considering waves on a rope.

Suppose you and a friend hold opposite ends of a rope and send wave pulses toward the middle of the rope, as in Figure 5.3. When those pulses meet, they're each telling the rope to do something (go up, go down, or stay still). Being an innocent bystander in this process, the rope does what both pulses tell it to do. If each pulse says "go up," then the rope rises twice as high

Figure 5.3

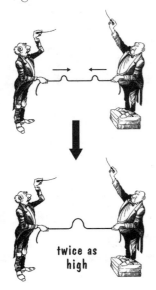
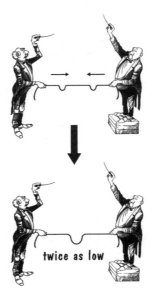
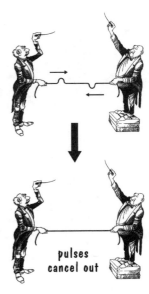

twice as high

twice as low

pulses cancel out

Figure 5.4

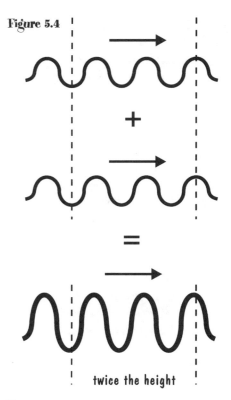

twice the height

Figure 5.5

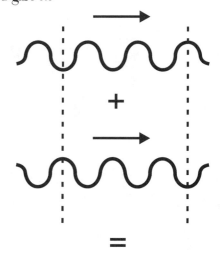

waves cancel and you get nothing

as it would with just one pulse. If each pulse says, "go down," then the rope goes down twice as much. If one pulse says to go up and the other says to go down, those instructions cancel each other, and the rope does nothing.

Sound waves do the same kind of thing. If waves from two separate sources hit a spot "in phase," meaning their up-and-down motions are in sync, then the total effect at that spot is twice what it would normally be (Figure 5.4).

If the two sources are "out of phase," meaning their up-and-down motions are totally out of sync, then the two waves cancel each other and you get nothing (Figure 5.5).

This process of waves adding to the largest possible amplitude, canceling to the smallest possible amplitude (maybe an amplitude of zero), or something in between, is known as **interference**. When two or more waves combine to produce a greater result than the individual waves, that's called **constructive interference**, and when two or more waves combine to produce a lesser result (they cancel one another) than the individual waves, that's called **destructive interference**.

Let's apply this idea to the sound waves coming from the two speakers. Figure 5.6 shows that the waves coming from each speaker usually travel different distances in getting to a given place.

In traveling different distances, the two sets of waves shift with respect to each other. If the total amount of shift is a wavelength, or multiples of a wavelength, the two sets of waves will still be in

Topic: interference

Go to: *www.scilinks.org*

Code: SFS10

Figure 5.6

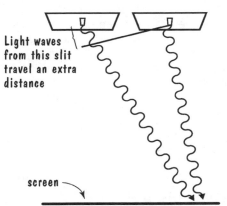

Waves from this speaker travel an extra distance

Light waves from this slit travel an extra distance

screen —

Just as with speakers, light reaching this point from one slit has traveled farther than light from the other slit

sync (in phase), and they'll interfere constructively. This gives you a loud spot. If the total amount of shift is a half a wavelength, or odd multiples of half a wavelength, the two sets of waves will be totally out of sync (out of phase), and they'll interfere destructively. This gives you a quiet spot. The differences are shown in Figure 5.7.

Overall, you end up with a pattern of loud-soft-loud-soft-loud-soft as you move around the speakers (Figure 5.8). This is due to the sound waves from the separate speakers getting in and out of phase because they travel different distances to get to you. For the record, let's call this a *spatial* pattern of loud and soft spots, because the pattern changes as you move through the space around you.

Figure 5.7

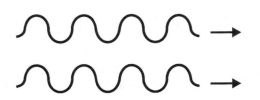 

Waves shifted a multiple of one wavelength are still "in sync"—a loud spot

Waves shifted an odd multiple of one-half wavelength are totally "out of sync"—a quiet spot

5 Chapter

Okay, let's move on to how pitch affects this pattern. If your hearing skills were on top of their game, then you noticed the following pattern. The higher the pitch of that annoying hum, the closer together the loud and soft spots in the pattern. The lower the pitch, the farther apart the loud and soft spots. And now that you know the "correct answer," you might want to go back to your radio and verify that I'm telling the truth.

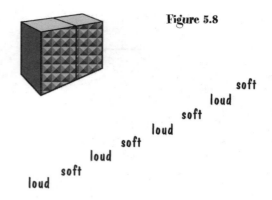

Figure 5.8

A teaching note: I'm sure all of us have had our personal experience contradicted by the "right answer" in the textbook. When trying to teach concepts to people, just about the worst thing you can do is rely on authority and tell the people what they *should* have observed. They observed what they observed, and that's that. If they didn't observe what you expected them to, then they might have done the procedure wrong, and it's time to have them repeat the procedure as you watch. Of course, since you and I are most likely very far away from each other, I can't observe what you do as you go through the activities in this book. All I can say is that I have done everything in the book, and consistently gotten certain results. If you *don't* get the results I describe, then chances are you aren't doing exactly as I asked, or possibly I just didn't explain the procedure well enough! Hey, just come up to me at my next workshop or book signing, and we'll figure it out!

Since this is an explanation section, how about an explanation? Let's take a closer look at what's going on with the pattern of loud and soft spots produced by two speakers. When the sound coming from one speaker is shifted by half a wavelength with respect to the sound from the other speaker, those two waves are totally "out of sync," and create a soft spot. If you move a bit further on, the sound from the two speakers will shift an entire wavelength with respect to each other, and they're back in step. This creates a loud spot. Figure 5.9 shows this.

Figure 5.9

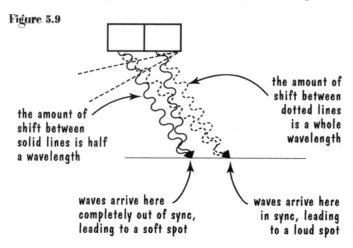

the amount of shift between solid lines is half a wavelength

the amount of shift between dotted lines is a whole wavelength

waves arrive here completely out of sync, leading to a soft spot

waves arrive here in sync, leading to a loud spot

Figure 5.10 shows this same situation for a higher frequency (smaller wavelength). With a smaller wavelength, the sound waves from the two speakers get in and out of sync in a shorter distance. That means the loud and soft spots are closer together with higher frequencies.

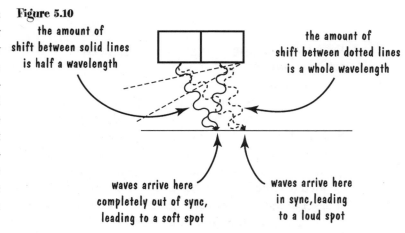

Figure 5.10

the amount of shift between solid lines is half a wavelength

the amount of shift between dotted lines is a whole wavelength

waves arrive here completely out of sync, leading to a soft spot

waves arrive here in sync, leading to a loud spot

About now you might be asking, "So what?" You'll have to wait for the applications section for that answer.

More things to do before you read more science stuff

Find a guitar, violin, or other stringed instrument. An electric guitar with an amp would be the best. Mess with the tension of two adjacent strings until those strings produce just about the same pitch when you pluck them. Then pluck them at the same time. In addition to hearing the notes they produce, you might hear a "wah-wah" effect, where the sound gets alternately loud and soft. If you don't hear that, alter the tension in the strings until you do. You'll know you're hearing the right thing when you find that the closer in pitch the two strings are, the slower the "wah-wah" effect. If you have an electric guitar, the "wah-wah" will be really easy to detect.

Figure 5.11

I know you're longing for the melodic sounds of a metal rod, so get those rods out and suspend one of them from the middle with a string. Make sure the string is tied tightly, because you're going to be spinning the rod, and it just wouldn't be fun to have the rod come loose and fly off and hit you in the face. Whack the suspended rod with the other metal rod. You've heard that sound before, so nothing new there. Next, hold the suspended rod in front of you and hit it so that it twirls in a horizontal circle after you hit it. See Figure 5.11. Hear the "wah-wah" sound? How does the "wah-wah" sound change when you change the speed at which the rod spins?

"The ol' triangle just doesn't sound the same since Junior flattened it."

More science stuff

Let's start with the two guitar strings plucked at the same time. You have two separate sources of sound, so at any point near the strings, you should hear a combination of those two sounds. The sound waves should add together, just as the sound waves from the two speakers of a boom box add together, right? Perhaps you remember what happened when sound waves from those two speakers added together (hey, it wasn't that long ago—just the first part of this chapter!). In some spots they interfered constructively (created a louder sound than that produced by each speaker alone) and in some spots they interfered destructively (tended to cancel each other and create a softer sound than that produced by each speaker alone). So, the same kind of thing should happen here, yes? Well, yes and no, actually. It's the same principle, but with a slightly different result.

Figure 5.12

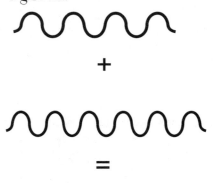

The key here is the frequencies of the sound coming from the two separate sources. With the boom box, the sound coming from each speaker was the *same frequency*.[2] With the plucked strings, you made sure each produced a slightly different note, meaning they produced *different frequencies*. Now, these different frequency sound waves still add and subtract, just as with sounds of the same frequency, but the result is a little bit different. This situation is shown in Figure 5.12.

Figure 5.13

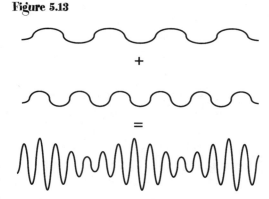

Okay, so what exactly is the result when you add these two waves of different frequency together? If you're feeling adventurous, you can go through point by point and see what happens when you add the waves together. What? You don't want to go through and add

---

[2] This might not be obvious, so let me explain. When you have a radio tuned to the AM dial (as you did when using the boom box), the radio sends out the exact same sound waves from each speaker. So, at any one time, the frequencies coming out of one speaker are pretty much identical to the frequencies coming out of the other speaker. Of course, when you have the radio tuned to FM stereo, you get different sounds coming from each speaker (that's what they mean by *stereo!*). Different sounds mean different frequencies.

and subtract the two waves point by point? Don't blame you, so you can take the easy way out and look at the result shown in Figure 5.13.

What you have in Figure 5.13 is a certain frequency of sound wave whose amplitude varies quite a bit. In some places it's just about zero, and in other places it's pretty large. Where the amplitude is small, you get a soft sound. Where the amplitude is large, you get a loud sound. In other words, you get a definite frequency of sound wave that alternately gets loud and soft, shown in Figure 5.14. Hey, that's the "wah-wah" effect! And just to distinguish this pattern from that in the earlier section, we'll call this a *temporal* sound pattern, meaning the alternate soft and loud sounds occur in time.

Figure 5.14

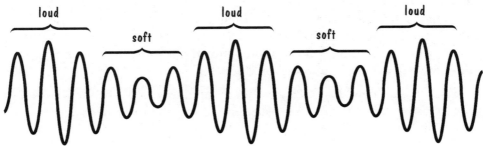

There's a special name for this "wah-wah" effect. It's known as **beats**. Whenever you add together two waves of slightly different frequency, you get beats. It turns out that the closer together in frequency the two sounds are, the farther apart the loud and soft spots in the beats. In other words, the closer together the two frequencies, the *slower* the beats. See Figure 5.15.

Figure 5.15

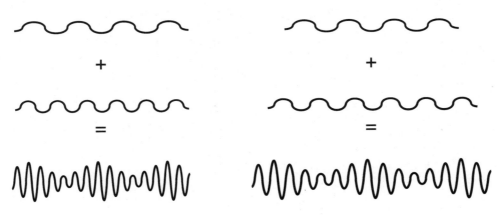

As the two frequencies get closer and closer to being the same, the beats become so slow that eventually you don't even notice them. If the two frequencies are exactly the same, there aren't any beats at all. Just for kicks, you might want to head back to your guitar and mess around with the beats you produced. By slowly changing the tension in one or both of the strings, you can control how fast or slow the beats are.

Okay, fine. We know what causes beats. It's two different sound sources that are almost the same frequency. So why in the heck do you hear beats when you hit a metal rod so that it's spinning? Isn't there just one source of sound (the rod)? Yes, but there's something else going on. Look back to Chapter 3 and review the Doppler effect. If you're too lazy to do that, I'll just go ahead and remind you of what that is. When a sound source is moving toward you, the sound waves in front of the source bunch up, creating a *higher* frequency sound wave than what the source produces. When a sound source moves away from you, the sound waves behind the source stretch out, creating a *lower* frequency sound wave than what the source produces.

With that in mind, let's take a look at the spinning rod. When it's in front of your face, one side of the rod is moving toward you and the other side is moving away from you (Figure 5.16).

Because of the Doppler effect, you hear a slightly higher frequency sound from the side moving *toward* you and a slightly lower frequency sound from the side moving *away* from you. So essentially, you have two sound sources, each producing a slightly different frequency. Hmmmm . . . beats! The faster the rod spins, the greater the difference between the two frequencies you hear. That means the beats are *faster* the faster the rod spins.

Figure 5.16

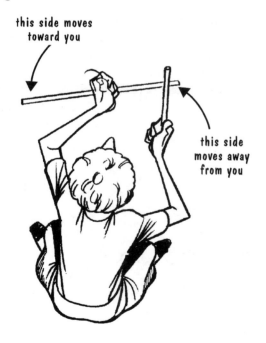

this side moves toward you

this side moves away from you

Chapter Summary

When different sound waves arrive at a given point, they tell the air molecules at that point to do different things (move one way or another with possibly different amplitudes). The air molecules will end up responding to the total effect of all the individual sound waves. When sound waves add or subtract, that's known as *interference*.

When the combination of two or more sound waves tells the air molecules at a given point to vibrate at a *greater* amplitude than would the individual sound waves, those waves are said to *interfere constructively*. When the combination of two or more sound waves tells the air molecules at a given point to vibrate at a *lower* amplitude than would the individual sound waves, those waves are said to *interfere destructively*.

When two sound sources of the same frequency add together, they produce a series of spatially distributed loud and soft spots.

When two sound sources of slightly different frequency add together, they produce a series of temporally distributed loud and soft spots (a "wah-wah" effect) known as *beats*. The closer together in frequency the two sounds, the slower the beats. The farther apart in frequency the two sounds, the faster the beats.

Applications

1. How sound is projected in a simple room or a concert hall is known as the *acoustics* of the room or hall. I used to think the best place to hear any kind of music would be a room that was completely lined with sound-absorbing curtains (or egg cartons!), so you wouldn't hear any echoes. Silly me. Suppose you had a concert hall that absorbed all sounds and didn't have any echoes. What would you hear? Well, the concert hall would probably have a whole bunch of "dead spots," where various sound frequencies added destructively. How well the band or symphony sounded would depend entirely on where you were sitting in the concert hall. Echoes to the rescue. If there are enough places for the sounds from the band to echo from, you hear particular notes not just from one place, but from several places. Because sound waves take time to travel different distances, those waves also hit you at slightly different times. Those slightly different times of arrival aren't enough to ruin the music you hear, but they are enough to "smooth out" the tendency for loud and soft spots to occur. The design of concert halls doesn't stop with just providing a place for echoes. If that were the case, all you would need for a good concert hall would be four walls and a

ceiling, all made of concrete. As I can attest from going to a concert in such a place, that doesn't cut it! The next time you're in a concert hall or just a plain ol' auditorium, look around you. There are curtains, yes, but there are also all sorts of small and large surfaces at many different angles, designed to produce the best sound for as many people in the hall as possible. Concert hall designers also need to be aware that the pattern of loud and soft spots is different for different frequencies, and that different frequencies reflect differently from different-sized surfaces. That's why the reflecting surfaces around the hall are of varying sizes, and it's also why the arrangement of amplifiers or musicians on stage can dramatically affect the sounds the audience hears.

Of course, the designers can get it wrong. I recall being at a particular brand new football stadium that contained a "state of the art" sound system. The music played over the sound system was totally garbled and you couldn't make out a single word the announcer was saying. The echoes, rather than enhancing one another, completely messed up the resulting sound. The following year, they relocated speakers and rethought the design, and you could hear perfectly. I would never reveal the name of this stadium, except to say that the professional football team that plays there rhymes with "froncos."

2. You can use beats to tune a guitar. When a guitar is properly in tune, each string produces a note that is directly related to the notes produced by all the other strings. For example, if you hold the top string of the guitar at the fifth fret and pluck it, you should hear the same note as when you pluck the string that's second from the top. So, to get the first two strings in tune, you hold the top string at the fifth fret and pluck it at the same time you pluck the second string. If the strings are in tune, you should hear the same note from each string. That's great, but our ears are fallible. How do we know the strings are producing the exact same note? The way we know is by listening to the beats. If you pluck the strings and hear really fast beats, you know the two notes produced aren't all that close in frequency. The thing to do is adjust the tension in the second string until the beats get slower and slower, and finally disappear. There are similar relationships between the notes produced by all the other strings, so you just continue from there. Of course, most guitarists these days use an electronic device to tune their guitars. But you know what? I think those devices rely on beats for their operation.

3. This application is pretty silly, but it's fun. You can create beats just by whistling with a friend. Find a friend who can whistle, and who can hold a steady note while doing so. Have your friend start to whistle with a constant note, and then you join in, trying to match the note. Chances are you won't

match the note exactly, and you'll end up with two slightly different frequencies, and a distinct "wah-wah" effect. Cool, huh?

4. You might think beats produced by a spinning rod are just a neat little trick, but there's something similar with light waves that gives us a lot of information about our Sun and planets. Because light travels in waves,[3] there is a Doppler effect for light. When the Sun or a planet spins, one side is moving away from you and one side is moving toward you. This motion alters the frequency of the light you see coming from the two sides. By comparing the frequency of light emitted from one side of the Sun (or a spinning planet), with the frequency of light emitted from another side of the Sun (or a spinning planet), you can determine the exact speed of rotation.

5. Some large amplifiers have a spinning mechanism with a speed control. Musicians use this spinning mechanism to produce a "wah-wah" effect in the emitted sound, just as a spinning rod produces a "wah-wah" effect. Bet most of those musicians don't know this is the Doppler effect in action!

---

[3] Waves constitute only one model of what light is. For more details, see the *Stop Faking It!* book on Light.

The Hills Are Alive...

Yep, you guessed it. This chapter is about music. Yes, we've already talked about music and musical instruments, but we're going to get into this subject in more detail. For example, you might already know that larger musical instruments tend to produce lower notes, but that isn't always true. A clarinet and a flute are about the same size, but the flute produces much higher notes. Why? Stay tuned (a little music humor there) and you'll find out. So, all together now, "Climb every mountain . . . ."

6Chapter

Things to do before you read the science stuff

What better way to learn about musical instruments than to get some string and a couple of metal washers? Trust me, it'll make sense later. So go ahead and get those materials. First cut a piece of string about 30 cm long and tie a washer to one end, as shown in Figure 6.1. This setup is known as a *pendulum.*

Holding the free end of the string, start the washer swinging. Notice how long it takes for the pendulum to swing back and forth each time. Note that this time doesn't change as the pendulum keeps swinging. In fact, there's a definite *frequency* (number of swings per period of time) that this pendulum seems to prefer. You could even call it a *resonant frequency.* No, this isn't a wave on a string, but it's still accurate to refer to a frequency that a pendulum "prefers" as a resonant frequency. Each time you set this particular pendulum swinging, it will move back and forth at its resonant frequency. For a pendulum such as the one you just built, it turns out that the length of the string is what determines the resonant frequency. To see this, just hold your pendulum halfway down the string and start it swinging. You should notice that this shorter pendulum has a much higher resonant frequency (it goes back and forth more times in a given time period).

Figure 6.1

Now build yourself a second pendulum that's the same length as the first. Check to see that these two pendulums have the same resonant frequency. Now tie both of the pendulums to a third piece of string, as shown in Figure 6.2.

Figure 6.2

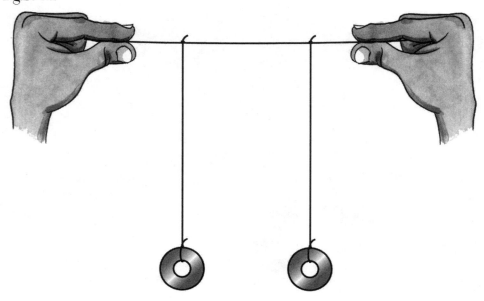

70

Now find a helper. While you hold the pendulums as shown in Figure 6.2, have your helper start one of the pendulums swinging. Figure 6.3 shows a top view of this.

Figure 6.3

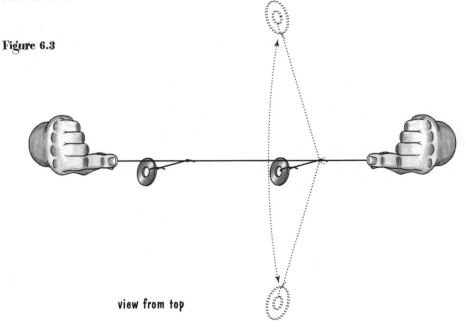

view from top

Watch and see what happens. Magic, huh?

Now shorten one of the pendulums so it's about half the length of the other one. Again, start one of the pendulums swinging and see what happens. No more magic.

Okay, time to move on to actual musical instruments. Get a guitar, violin, or other stringed instrument. Adjust the tension in two adjacent strings until those two strings produce the same note when plucked. If you understood what was going on in the previous chapter, you can use *beats* to make sure the two strings are producing the same note. Pluck one of the two strings and then put your finger on that string so it stops vibrating. Do you still hear the note? To find out what's making that sound, put your finger on the second string. Try this a few times until you're convinced you can cause a second string to vibrate just by plucking the first string. For extra credit, figure out what the two guitar strings and the two pendulums have in common. If you're not interested in extra credit, just read the next section for all the answers.

The science stuff

Remember that resonant frequencies are frequencies that a string or a pendulum *likes*. If you try to make an object vibrate at its resonant frequency, the object will respond strongly. Let's see how that applies to the two pendulums tied to a string. When the pendulums are the same length, they have the same resonant frequency. When you start one of the pendulums swinging at that resonant frequency, the frequency of the movement is transmitted through the top string to the other pendulum. The other pendulum responds well to this frequency, so it starts swinging. In fact, the response of the second pendulum is so strong that the entire swinging motion transfers from the first to the second pendulum. The reason the first pendulum stops swinging has to do with something called *conservation of energy*, which is covered in the *Stop Faking It!* book on Energy. When you shortened one of the pendulums to half its original length, you changed that pendulum's resonant frequency. With different resonant frequencies, the swinging motion of one pendulum doesn't transfer very well to the other pendulum.

Maybe by now you've figured out what's going on with the guitar strings. If not, here's the explanation. When you adjust the tension in the two strings so they play the same note when plucked, you are adjusting them so they produce the same resonant frequencies when plucked. They will also *respond* strongly to those resonant frequencies. So, you pluck one string and then put your finger on it to stop its vibration. By that time, however, the second string has responded strongly to the sound produced by the first string, and the second string is now vibrating and producing that same note. If you happen to have a set of tuning forks around (doesn't everyone?), you can get the same effect using two tuning forks that produce the same note. Strike one and hold it near the second. Then stop the first one from vibrating, and the second one will be producing the note because it responds strongly to the resonant frequency.

Okay, what does all this have to do with instruments? Lots. First, I have to tell you that I lied to you just a wee bit back in Chapter 3. There I explained that plucking a string caused the air around it to vibrate at the same frequency as the string, and these air vibrations were the sound waves you hear. Although a vibrating string does cause the air around it to vibrate, that's not the whole story. A plucked string moves very little air, and that small amount of vibrating air can't account for all the sound you hear from a guitar or a violin. It turns out that the body of a guitar is constructed so it will respond strongly to a wide range of frequencies. So when you pluck a string on a guitar, that vibration causes the body of the guitar itself to vibrate. The motion of the guitar body moves lots of air, and you hear all that nice sound.

Armed with this information about guitars, one can understand how a poorly-constructed guitar body might do a lousy job of resonating when you pluck the strings. One can also understand how it might not be all that easy to construct a guitar body that resonates over a wide range of frequencies. And one might understand why some guitars cost $50 while others cost $1,000!

Topic: resonance

Go to: *www.scilinks.org*

Code: SFS11

The situation gets even trickier with violins. How well a violin body resonates in response to the strings being bowed is critical for the quality of the sound produced. Some violinists claim that even the type of varnish used on the violin body can significantly affect the sound of the violin. Personally, my hearing has never been quite that discriminating! If you think for a while, you can probably come up with many other instruments besides guitars and violins that rely on resonance to produce nice, full sounds. Here are a few to get you started: pianos, drums, mandolins, sitars, cellos, kalimbas, and xylophones.

More things to do before you read more science stuff

I haven't sent you to the hardware store in a while, and I'm sure you miss the place, so I'm going to ask you to buy a short length (maybe 30 cm long) of hollow copper tubing. It should be at least 1/4 of an inch in diameter and no wider than about 3/4 of an inch in diameter. With a little practice, you should be able to produce a sound by blowing across one of the open ends of the tubing, as shown in Figure 6.4.

Once you do that, close one end with your finger and blow again. This should produce a lower note. If you are a music person, see if you can figure out the relationship between the two notes you produce. If you have a recorder around the house, cover all the holes and blow, producing the lowest note possible on a recorder. Then repeat, placing the bottom opening against your knee so it's completely covered. Hmmm . . . guess that first note *wasn't* the lowest possible note!

Figure 6.4

As a final "thing to do," think about all the different musical instruments you might find in an orchestra. What sounds does each instrument produce? What's the relationship between the size of the instrument and the sound it produces? Do you suppose there is a purposeful relationship between the sizes of different instruments?

More science stuff

The obvious connection between the size of a musical instrument and the sound it produces is that bigger, longer instruments produce lower pitches. Compare the sounds produced by a piccolo (small) and a flute (large), an oboe (small) and a bassoon (large), and a trumpet (small) and a trombone (large). Of course, we already know why longer instruments should produce lower sounds. Whether we're talking about standing transverse waves on a string or standing longitudinal sound waves inside a wind instrument, the longer the instrument, the longer the wavelength of the standing waves produced. Longer wavelengths mean lower frequencies, so longer instruments tend to produce lower notes.

Okay, if that's true, then what about the question I posed at the beginning of this chapter? How can a clarinet, which is just about the same size as a flute, produce much lower notes than a flute? To answer that, let's look at what happened with your hollow copper tube. When you blow across the top of the tube, you produce a whole set of standing waves, including the fundamental and all the harmonics. Let's just look at the fundamental, keeping in mind that because both ends of the tube are open, those ends must be antinodes.

Figure 6.5

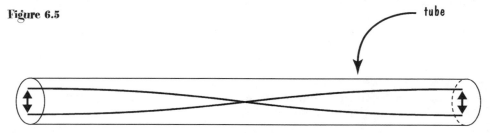

Remember that in drawings such as Figure 6.5, you are looking at a "continuous motion" drawing. If you look at just a snapshot of the fundamental standing wave, it looks like Figure 6.6.

Figure 6.6

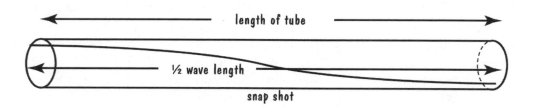

From Figure 6.6, you can see that the length of the tube is equal to one half a wavelength. Now let's look at what happens when you close one end of the tube. When you do that, you force the closed end of the tube to be a node rather than an antinode. Now the fundamental standing wave (both continuous motion and snapshot) looks like Figure 6.7.

Figure 6.7

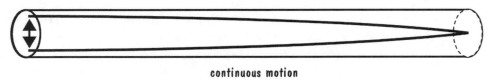

continuous motion

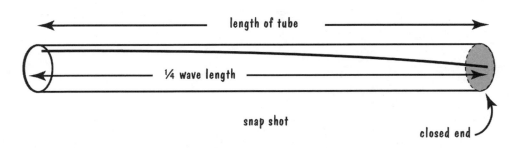

length of tube

¼ wave length

snap shot

closed end

With one end of the tube closed, the length of the tube is now equal to one fourth of a wavelength. In other words, by closing one end of the tube, you double the wavelength of the fundamental standing wave. Doubling the wavelength means the frequency of the fundamental is exactly half of what it was with both ends open.[1] In musical jargon, when you change the frequency by a factor of 2 (doubling it or halving it), that's known as changing the frequency by one **octave**. The curious thing about two notes that differ by an octave is that they sound like the same note to us, with one just being higher or lower. To experience this, find a piano and play all of the C notes. These all differ by an octave (a doubling of frequency), and they all sound like the same note to us.

So, you can change the fundamental frequency of a copper tube by one octave just by closing one end. You can also do the same with a recorder. And, you can do the same with a clarinet. The clarinet is a reed instrument, which

---

[1] By now, you should be able to use $f\lambda = v$ to show this. Just remember that v stays constant.

means you create standing waves inside the clarinet by causing the reed to vibrate. When that reed vibrates, it effectively makes that end of the clarinet a closed end. A flute, on the other hand, is open at both ends. Because the clarinet is closed at one end, it can produce notes roughly one octave lower than those produced by a flute, even though the two instruments are basically the same size.

I asked you in the previous section whether you thought there was a purposeful relationship between the sizes of the different instruments in an orchestra. The answer is yes. Because you want all the sounds of an orchestra to fit together in a way that is pleasing to the ear, it's a good idea for different instruments to be related according to octaves. If you double the size of an instrument, you halve its fundamental frequency, and the two lengths will result in notes that differ by an octave. Hence, a trombone is twice the length of a trumpet. A baritone sax is twice the length of a regular saxophone. Of course, these differences in length don't have to be exact, because you can compensate somewhat by where you put the various holes or frets in an instrument, but you'll find that the length relationship is generally true throughout an orchestra.

Chapter Summary

When you stimulate any object with a vibration that matches the object's resonant frequency, the object responds strongly to that frequency.

If the resonant frequencies of two sound sources are identical, one source of sound can activate a second source of sound without actually touching that second source of sound.

Resonance allows the body of a musical instrument to increase the volume and fullness of the sounds produced by the instrument.

The size of a musical instrument affects the range of pitches produced by that instrument.

Whether a wind instrument is open at both ends or closed at both ends also affects the range of pitches produced by the instrument.

An octave is an exact doubling or halving of a frequency.

Musical instruments are designed so the notes they produce differ from the notes produced by other instruments by an integral number of octaves.

Applications

1. Maybe you're old enough to remember that Memorex commercial where Ella Fitzgerald breaks a wineglass with her voice. That's possible to do, if you know a little physics. When you hit a wineglass, it will vibrate at its resonant frequency. This vibration is actually two dimensional, where parts of the glass rim move inward at the same time other parts move outward. See Figure 6.8.

   If you sing a note with a frequency that exactly matches the resonant frequency of the glass, the glass responds strongly to that note. In fact, the vibrations of the glass can get so large that the glass shatters.

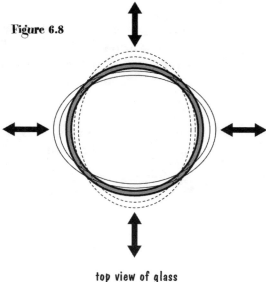

Figure 6.8

top view of glass

2. Have you ever gotten a little queasy just being at an airport? That queasiness might not be anxiety about flying, but rather the phenomenon of resonance. Some of the low frequency sounds produced by jet engines match the natural resonant frequency of your internal organs. Your insides start vibrating in response to those jet noises, and that makes you feel not-so-well. And for you fans of *Dune*, maybe now you can figure out the science behind "weirding modules" blowing people apart with sound!

3. Are you familiar with what is called "feedback" on electric guitars? If not, scrounge up some music by Jimi Hendrix, The Grateful Dead, or any other band that played about the same time as those two. What happens in feedback is that the guitarist plays a note and then puts his or her guitar near the amplifier. This results in a sustained, almost annoying sound that's more of an electrical hum than it is music.[2] What happens is that the frequency coming out of the amplifier speaker causes the guitar pickup to vibrate even more than it was before (resonance). This in turn increases the volume coming out of the amplifier, which increases the vibration of the guitar pickup, which increases . . . . In other words, you have runaway resonance. By the way, you can get a much more annoying form of feedback just by placing a microphone too close to its amplifier (see one of the early scenes in the movie *Toy Story!*). The source of that squeal is the same as the source of guitar feedback.

---

[2] Although in the hands of the right guitarist, feedback can sound pretty nice.

4. We know it's possible to produce sounds on guitars and violins by plucking the strings, but we also know that the most common way of producing sound on a violin is with a bow. How does a bow produce notes when you draw it across violin strings? Believe it or not, it's the same as what happens when chalk squeaks on a chalkboard (see Chapter 1 Applications). The bow pulls the violin string a bit to the side and then the string slips back into place, only to be pulled aside again by the bow, and only to slip back in place again. This sets up vibrations in the violin string. Of course, there's a definite art to coordinating this pulling and slipping so the sound you get is pleasant rather than something that sounds like squeaking chalk!

Listening Devices

We've covered most of the basics of how sound is produced and transmitted, but we've left something out, and that's all the electronic devices we use to amplify sound, reproduce it, or get it from one place to another. How exactly does a CD player work? How does a phone work? And how exactly do those two listening devices we carry on either side of our head work? No, not headphones, even though it might seem that some kids do have those permanently attached. I'm talking about ears.

Things to do before you read the science stuff

Find yourself an old person—you know, someone older than 40. If you happen to be older than 40 yourself, you'll do. Chances are, that old person has what is known as a phonograph turntable, or record player. Along with a record player,

"I heard that! And don't think I won't remember it."

the old person might also have what are known as records. These are large discs made of vinyl. For all you young people, we old people actually used records and record players to play music. Hard to believe, I know.

Once you find a record, take a close look at it. How in the world do you suppose music is stored on this thing? You might get a hint by using a magnifying glass and looking closely. Now take a look at the record player. It has a needle attached to an arm. Why would placing a needle on the record produce sound? To show that placing a needle on a turning record actually does produce sound, do the following activity. Find a really old record by an artist you are sure you never want to hear again.[1] Form a sheet of paper into a cone and tape it, leaving the tiniest of holes at the tip of the cone. Drop a straight pin into the cone so the sharp end sticks out the tip of the cone. If the pin falls all the way through, the hole at the tip of the cone is too large. Try again until you get something like Figure 7.1.

Figure 7.1

Now place your unwanted record on the turntable and start it spinning. Hold the cone and pin above the record so the point of the pin just touches the record, as in Figure 7.2. This should prove to you that simply placing a sharp object on a record produces music.

Let's move into modern times and take a look at a CD. Other than the fact that it's round, does a CD look anything like a record? There's music stored on this thing, right? The question is *how*. Here's a hint: Notice how a CD is almost mirror like, and that it makes pretty rainbows when you shine light on it. Not much of a hint, is it?

Figure 7.2

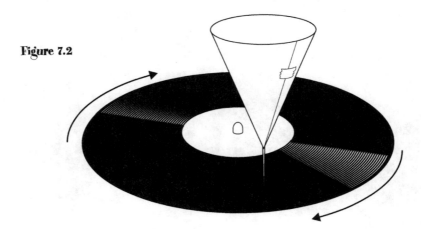

---

[1] You will likely ruin this record. My wife has definite candidates for this activity from my record collection, but she's wrong about my album by *The Flock*.

History time again. See if you can find an old phone with a handset shaped as in Figure 7.3. With this type of phone, you can unscrew the cover from both ends—the part you listen to and the part you talk into.

Figure 7.3

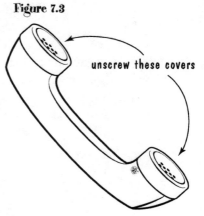

unscrew these covers

Go ahead and unscrew those covers and see what's inside.[2] What you'll find are what look like mini-speakers with magnets attached. Don't believe that's a magnet? Bring a paper clip near it and see if the paper clip is attracted.

Now find yourself a stereo speaker that you can take apart. If you don't feel like taking apart your stereo, just head to a local electronics or auto store and look at replacement speakers for cars, the ones that are just the speakers without any casing. Look familiar? Yep, it's just a larger version of what's inside a phone, complete with magnet.

The science stuff

Let's start with speakers that are inside stereos and phones. Why is there a magnet attached? To understand that, I'm going to tell you about a very strange thing that happens when magnets and wires get together. When you move a magnet near a wire, that causes an electrical current to flow in the wire. That electrical current, being caused by the motion of the magnet, contains all the information about the exact motion of the magnet. The reverse is also true. If you run a changing (varying in amplitude or direction) electrical current through a wire, that changing electrical current will cause a nearby magnet to move. The motion of the magnet depends entirely on the exact nature of the changing electrical current in the wire. There is a rather involved activity you can do to convince yourself that I'm telling you the truth about electrical currents and magnets, and that activity is in the *Stop Faking It! Energy* book. I just didn't see the need to repeat that activity here.

Now we're ready to understand how phones work. You speak into the bottom part of a phone. The sound waves you produce cause the magnet inside the phone to vibrate. That magnet vibration causes an electrical current to flow in wires that surround the magnet, and that electrical current contains all the vibration information that you produced. The electrical current is transmitted

---

[2] Modern phones have basically the same mechanism inside that old phones do. The only difference is that modern phones are a real pain to take apart.

to someone else's phone.[3] When that electrical current reaches the other phone, it causes the magnet in the top part of that phone to vibrate and move the tiny speaker that's attached to the magnet. This reproduces the sound waves that started the whole process when you spoke into the phone on your end. So, the process isn't really all that complicated. On your end, your phone changes sound waves into electrical signals. On the other end, those signals change back into sound waves. Take a look at Figure 7.4.

Figure 7.4

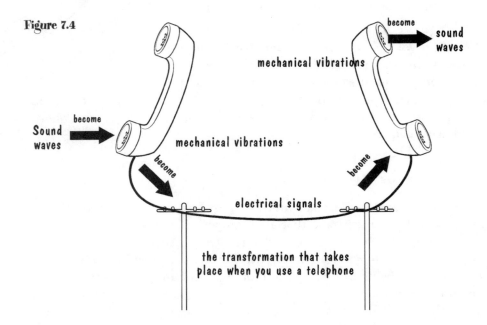

mechanical vibrations

become — sound waves

Sound waves — become — mechanical vibrations

become — electrical signals — become

the transformation that takes place when you use a telephone

Of course, this transmission process isn't 100 percent efficient. Some of the electrical information gets lost along the way, which is why the voice you hear over a phone doesn't sound as if the person were right next to you. I should also mention that there are other mechanisms besides magnets to turn vibrations into electrical signals and vice versa, so not all phones use magnets. Certain kinds of crystals produce an electrical current when they're deformed (as in when they vibrate), and this can actually produce much clearer sound transmission that you can achieve with magnets.

SCiLINKS.
THE WORLD'S A CLICK AWAY

Topic: telephone technology

Go to: *www.scilinks.org*

Code: SFS12

---

[3] It used to be that this electrical current was transmitted strictly by wires that physically connected to another phone. Nowadays, the transmission is accomplished at least partially with microwave signals or fiber optic signals. See the *Stop Faking It!* book on Light for how that might work.

Now we're ready to understand phonographs. A close-up view of a record (Figure 7.5) shows that the record has one continuous groove, which the needle follows as the record turns. This groove isn't smooth, though. The sides are uneven, and this causes the needle to vibrate back and forth.

Also, the sides are uneven in such a way that they cause the needle to vibrate in accordance with the original sound waves that were used to create the record.

Your evidence of this is the fact that your straight pin and paper cone combination produced music. The paper cone merely amplifies the vibrations of the straight pin. Of course, a stereo doesn't use a paper cone to magnify sound waves (although early record players—think of the RCA dog listening to "his master's voice"—used a metal cone). Instead of a cone, phonographs reproduce sound almost exactly as phones do. The grooves in the record cause the needle (referred to as the *stylus*) to vibrate. This vibration is transformed into electrical signals, which are transmitted to the speaker, where they are transformed back into vibrations (using the magnet in the speaker) that produce sound waves. Before the electrical signals reach the speaker, however, they are *amplified* electronically so the sound coming out of the speaker is much louder than you'd get if you just attached the needle to the speaker cone.

Figure 7.5

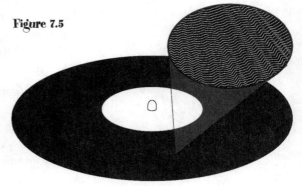

close-up view of the grooves in a record

And now on to CDs. No matter how closely you look, you can't see any grooves on a CD. Even though you can't see them, they're there! The grooves on a CD are much smaller than those on a record, and they're smaller not because the CD is smaller than a record, but because the grooves on a CD are designed to reflect light waves. And in fact, that's what they do. A CD player shines a laser beam onto the CD, and the CD reflects the laser beam in a way that corresponds to the sound signals that were used to create the CD. The reflected laser signals are transformed into electrical signals (the mechanism for that is a bit beyond the scope of this book), and those electrical signals in turn, after being amplified, cause the magnet in a speaker to move. Not so complicated, really.

Almost every electronic device that amplifies or transmits sound works in pretty much the same way. Vibrations become electrical signals, which then become vibrations. That's how microphones, electric guitars, and PA systems work. Of course, there are exceptions. Electric pianos and synthesizers bypass the initial vibrations. They start with electrical signals and transform them into vibrations.

More things to do before you read more science stuff

Time to make another cone out of paper. This time, however, make the opening at the tip of the cone a few centimeters in diameter. What you're going to do with this cone is pretty simple. First, don't use it at all (told you it was simple). Listen to

Figure 7.6

a radio, the television, or someone talking, with one of your ears directed toward the source of sound. Next hold the cone up to your ear (don't shove it in your ear unless you want to take a trip to the emergency room) and listen to the source of sound again (Figure 7.6). Notice any difference? You should.

Next get a ruler (wooden or really stiff plastic), a pencil, and a rock or other heavy object. Lift the rock directly to get a feel for how heavy it is. Then set up the ruler, pencil, and rock as shown in Figure 7.7. The pencil should be in the middle of the ruler.

Figure 7.7

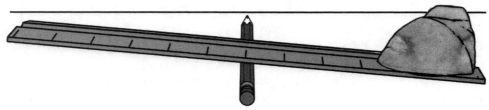

Push down on the free end of the ruler. This should raise the rock up. Notice how hard you have to push on the ruler in order to get the rock to go up. Also notice how high the rock rises, as in Figure 7.8.

Reposition the pencil so it's closer to the end with the rock, as shown in Figure 7.9. Again push down on the free end of the ruler to raise the rock. How

Figure 7.8

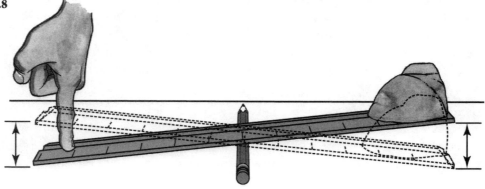

hard do you have to push compared to the setup in Figure 7.7? Also notice how high the rock rises compared to the setup in Figure 7.7. Amuse yourself by guessing what in the world a ruler and rock have to do with how your ear works.

Figure 7.9

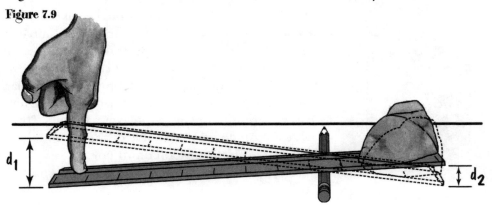

Get a widemouthed jar and a pop or beer bottle with a small opening (both empty and both made of glass). Also get those pieces of balloon you used earlier, along with the rubber bands. Fill the jar and the bottle all the way to the top with water. They both should be overflowing with water, so there's no air gap between the water line and the top of the container (Figure 7.10). Yeah, it might be a good idea to do this over a sink.

Figure 7.10

Now stretch one piece of balloon completely over the top of the widemouthed jar and secure it with a rubber band (Figure 7.11). It's best to have a helper for this, because you want to make sure there are no air gaps between the balloon and the water. If you do this correctly, water should spill over the sides of the jar as you're putting the balloon in place. And yes, the sink is still a good idea.

Figure 7.11

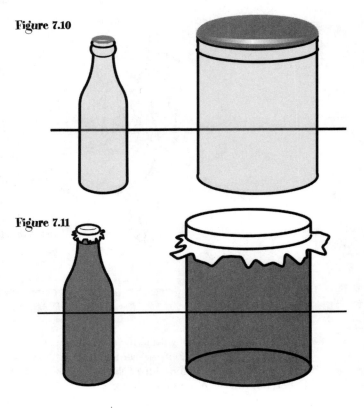

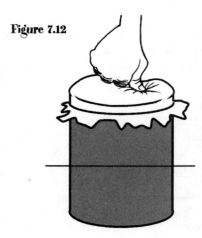

Figure 7.12

Do the same with the other piece of balloon and the pop or beer bottle. Again make sure there are no air pockets between the balloon and the surface of the water.

Use your thumb to push down on the balloon piece that's over the beer bottle. Make sure your thumb covers the entire opening as you do this. How much can you compress the balloon? Not much, huh?[4] Try the same thing with the piece of balloon covering the jar, like in Figure 7.12. This time, of course, it's impossible to cover the entire opening with your thumb. That's okay. Just push down on one side of the balloon. How far can you push the balloon down now?

More science stuff

Figure 7.13 shows a drawing of your basic ear.

Figure 7.13

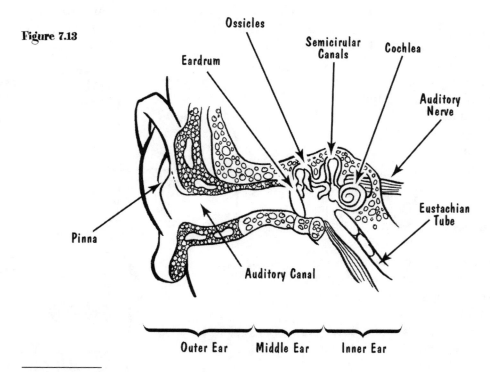

---

[4] If you push hard enough, and the rubber band isn't all that tight, water will escape between the balloon and the edge of the bottle. This will allow you to push down on the balloon. If this happens, redo the setup with a tighter rubber band. With a tight enough rubber band, you won't be able to push down on the balloon.

You've undoubtedly seen this kind of diagram before, and maybe you even memorized all the parts of the ear and what they do. I'm betting, however, that you probably didn't *understand* how everything worked. Otherwise, you'd be skipping this section! Anyway, to understand how the ear works, it helps to unravel the complex tangle of parts and instead use a simplified diagram, such as the one in Figure 7.14. Basically what I've done there is line everything up and straighten out the parts of the middle ear and the inner ear. Keep in mind that Figure 7.14 is *not* what your ear actually looks like, but rather is a stretched-out picture that helps us understand what's going on. You'll probably want to refer back to Figures 7.13 and 7.14 as you make your way through the rest of this section.

Figure 7.14

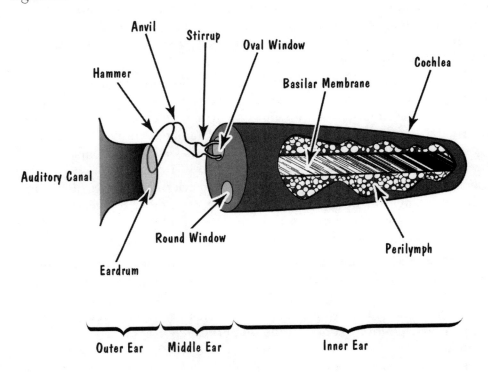

Let's start from the outside and move inward. The first part of the outer ear is the **pinna**. That's the part of your ear that you can see (providing you're looking in the mirror). It's pretty easy to guess what this does, but in case it's not obvious to you, think back to what happened when you used a cone of paper to listen to a sound source. With the cone in place, the sound got louder. The

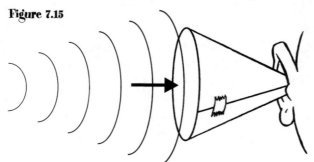

reason for that is that the paper cone gathered in more sound waves than your ear could. The cone intercepts a larger area of sound waves and reflects them toward your ear, as shown in Figure 7.15.

Figure 7.15

The pinna of the ear is shaped a bit like a cone, and its purpose is to gather in more sound waves than a simple little hole in the side of your head could. Would you be able to hear better if your pinna were shaped like a paper cone? Sure, but you'd have trouble getting dates.

After the pinna comes the **auditory canal**. It stands to reason that the purpose of the auditory canal is to protect the eardrum from dirt, Q-tips, fingers, and pencils.[5] Ruptured eardrums would be pretty common if the eardrum were right there at the surface of your head.

This brings us to the eardrum, or as people who want to sound important call it, the **tympanic membrane**. All the eardrum does is respond to incoming sound waves, transmitting vibrations of air molecules to the other parts of the ear. The eardrum is a pretty special piece of equipment. Recall that all objects have certain resonant frequencies to which they respond strongly. Clearly, the eardrum has to respond to an entire range of frequencies, and it has to respond just about the same to all those frequencies. If it didn't, then some frequencies would be a whole lot louder for us than other frequencies. To be honest, I'm not sure how the eardrum manages to respond equally to all frequencies we can hear. Must be magic.

Past the eardrum, you get to the middle ear. This consists of three bones, called the **hammer**, the **anvil**, and the **stirrup**. Those are the technical names. The everyday names are the *malleus*, the *incus*, and the *stapes*. Oh wait, those are the everyday names if all you speak is Latin. Whatever you call them individually, they are collectively known as the **ossicles**. Yep, *that's* easy to remember! To understand what the ossicles bones do, we have to go back to the ruler, pencil, and rock activity you did. What you should have discovered there was that, with the pencil in the middle of the ruler, in order to lift the rock up, you had to push down on the ruler just about as hard as you had to lift in order to

[5] This last example stems from the time I took my son to the emergency room because he had broken off the tip of a pencil inside his ear. I'm sure he'll appreciate me mentioning this if he ever takes the opportunity to read this book.

lift the rock directly.[6] When you move the pencil closer to the rock, you don't have to push nearly as hard in order to lift the rock. However, you should also notice that the rock doesn't rise as high as it does with the pencil in the middle. In other words, you got a trade-off. You didn't have to push as hard, but the rock didn't rise as far. A device like your ruler and pencil is known as a *lever*. A lever can get you an increase in force at the expense of a loss in distance moved.[7]

The ossicles bones, in combination, act like a lever. They're connected at one end to the eardrum, and at the other end to something called the **oval window**. The oval window is a membrane that's not a whole lot different from the eardrum, except that it's smaller. The eardrum pushes on the hammer bone with a certain force. Because the ossicles bones act like a lever, this causes the stirrup bone to exert a *larger* force on the oval window. The magnification isn't much—maybe one and a half times—but aided by the fact that the oval window is smaller than the eardrum, the whole mechanism can exert a large pressure on the inner ear, much larger than the incoming sound waves could.[8]

Maybe right about now you're thinking that magnifying the effect of incoming sound waves could be dangerous. What if those incoming waves are pretty gosh darned loud in the first place? Won't magnifying them hurt your inner ear? The answer is yes. There is, however, a defense mechanism. Loud sounds trigger, through a reflex action, the operation of two different muscles. One is known as the *tensor tympani* (gotta love those Latin names). This muscle alters the tension in the eardrum so the eardrum doesn't move quite as much when the incoming sound waves have a large amplitude. The other is known as the *stapedius*. This muscle limits the motion of the ossicles. This would be like restricting the motion of the ruler when you're trying to lift the rock. That makes the ruler, and in fact the ossicles, less efficient as a lever. The force mag-

---

[6] I realize that's a tough comparison to make, if only because in one case you're pushing and in the other you're pulling. You could go back and compare the two by performing the activity a bunch of times, or you could just take my word for it that the push in one case and the pull in the other are about the same.

---

[7] For more than you ever wanted to know about levers, check out the *Stop Faking It!* book on Energy.

---

[8] When you're doing science, you usually have to be really careful about terms such as *force* and *pressure*. Those are two very different concepts, and they shouldn't be used interchangeably. It's not worth the space right here to distinguish between the two, because as long as you get the general idea that the middle ear manages to increase the effect of incoming sound waves, that's all that's necessary. Of course, you can trust that while I'm not explaining the difference between these two terms, I *am* using them correctly. And if you want a thorough explanation, look for the upcoming *Stop Faking It!* book on Water, Air, and Weather.

SCi LINKS.
THE WORLD'S A CLICK AWAY

Topic: the ear

Go to: www.scilinks.org

Code: SFS13

nification is reduced, protecting your inner ear from loud sounds. As a side note, I just have to say that every time I learn about various body systems, it just amazes me how humans are put together. There are neat little mechanisms to accomplish all sorts of things, and then there are safety mechanisms and backup mechanisms to support those. There seem to be mistakes (the appendix!), but all and all the body is pretty remarkable.

Okay, enough fawning over the human body. Let's move on to the inner ear. There are three *semicircular canals* that have a cool way of keeping you balanced but have nothing to do with your hearing. Then there's the **cochlea**—a long coiled tube (a straight tube in Figure 7.14) that's filled with a fluid called **perilymph**.[9] As we'll see later, the motion of this perilymph is important for hearing, so it would be nice if pressure applied to the oval window would cause this fluid to move. At this point, I'd like to call your attention to the activity with the widemouthed jar and bottle that were covered with balloon sections. When you push down on the top of the bottle while covering the entire opening, you don't get very far. That's because most fluids, such as water, are *incompressible*, which basically means that they don't "give" when you push on them. Because the bottle is made of glass (which also doesn't "give") and you're completely covering the opening, there's nowhere for the water inside to go, so it's just like pushing on something solid. The result is that you can't get the water inside to move with this setup unless you push so hard the water squirts out between the balloon and the bottle.

this part moves up, allowing water inside to move

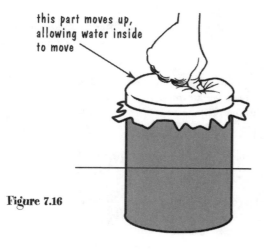

Figure 7.16

Now on to the widemouthed jar. When you push on one part of the balloon, you still have a whole bunch of nice, stretchy balloon left over for the water inside to push on. Because the balloon does stretch, there's a place for the water to go when you push down, as shown in Figure 7.16. It's easy to cause the water inside to move with this setup.

Okay, back to the inner ear. If the only membrane connected to the cochlea is the oval window, then you have a situation similar to the soda pop bottle. You can push on that oval window all you want, and the fluid

[9] You can always tell when you get to the subject of biology, because the number of vocabulary words increases exponentially!

won't move much, if at all, because there's no place for it to go. Tra-la-laaaaa![10] The **round window** to the rescue! The round window is nothing but another membrane connected to the cochlea and nothing else, but it's important. It acts just like the "rest of the balloon" on the widemouthed jar. It can stretch, allowing the fluid inside the cochlea someplace to go when pressure is applied to the oval window. Because the fluid inside has someplace to go, it can move around.

But why do we want the perilymph (that's the fluid inside the cochlea) to move around? Glad you asked. I won't go into all the details of the structure of the cochlea, but there's something called the **basilar membrane** (more vocabulary!) that runs along the center of the cochlea (Figure 7.17). Along the basilar membrane are twenty to thirty thousand tiny fibers. The fibers close to the oval window are short and stiff. The fibers gradually get longer and flimsier as you move toward the end of the cochlea. Because each fiber is a different size and shape, each fiber has a different resonant frequency. When a given frequency wave comes cruising along through the perilymph, only the fibers that have a resonant frequency matching that frequency respond strongly. That means only a small portion of the basilar membrane is stimulated by an incoming set of waves.

In case you miss it, I want to point out how great it is that basic ideas about how length and tension affect resonant frequencies apply to something as complex as the ear. Different length fibers and different tensions in the fibers affect how the fibers respond to incoming frequencies. Convenient, isn't it?

There's something called the **organ of corti** that lies along the surface of the basilar membrane. The organ of corti contains thousands of tiny little hair cells, which respond to the motion of the fibers in the basilar membrane.

The tiny hair cells stimulate something called the **auditory** (or **cochlear**) **nerve**, which carries electrical signals to the brain. Because only a particular set of hair cells are stimulated by a particular set of fibers, not only does the auditory nerve know that sound waves are hitting the outer ear, it knows what frequencies are hitting it. The brain interprets all the signals coming from the auditory nerve as various sounds. Cool, huh? While you're thinking that's cool, you might also notice that the whole operation of the ear isn't that much different from the operation of other listening devices. Vibrations cause other vibrations, which eventually transform into electrical signals (the nerve impulses).

---

[10] Lest we get a lawsuit, let it be known that I stole this from the children's book *Captain Underpants*. Really great stuff for all us children out there, but I do suspect it's "guy humor."

Figure 7.17

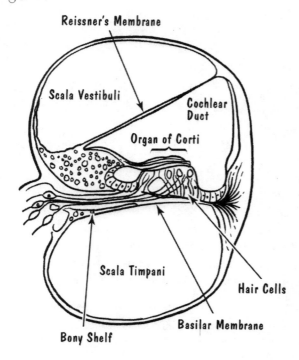

Reissner's Membrane

Scala Vestibuli

Cochlear Duct

Organ of Corti

Scala Timpani

Hair Cells

Basilar Membrane

Bony Shelf

There's one part of Figure 7.13 I haven't mentioned, so I guess I'll mention it. The **eustachian tube** leads away from your eardrum and down into your sinus cavities and your throat. The purpose of the eustachian tube is to help equalize pressure on the inside and outside of your eardrum. If you didn't have this tube, large increases or decreases in pressure (such as happen when you ride in an airplane or go underwater) would rupture your eardrum. If your ears are plugged, yawning usually helps because that allows outside air into the eustachian tube, helping to equalize the pressure on either side of the eardrum. If you've ever ridden in an airplane with a sinus infection that plugs up the eustachian tube, you have an appreciation for what it does!

Before finishing up on the ear, I should mention that we still don't fully understand how the combination of your ear and brain figures out what sounds are entering the ear. For example, the loudness of a sound (how big the amplitude of the sound wave is) and the pitch of a sound (its frequency) seem to affect each other. Changes in loudness can actually affect the pitch that people hear, even though the frequency as measured by instruments is unchanged. People also perceive differences in loudness between two different pitches that have the same loudness as measured by the amplitude of the sound wave. To add to the confusion, there's a term *timbre*, which refers to the quality of the sound we hear (screeching chalk as opposed to well-played music, for example). Some of our perception of timbre can be explained by analyzing the exact sound waves that different sources produce, but that doesn't explain everything.

Well, that's about it for sound. As with all the books in this series, I haven't explained all there is to know about sound.[11] If I did that, you'd have one of

---

[11] Two things I haven't covered in this chapter are the ranges of sound frequencies people can hear and the ways we measure the loudness of different sounds. You can find a discussion of these topics in just about any book on sound, but just in case you want a quick overview, check out the Applications section.

those intimidating textbooks in your hands. I'm just trying to arm you with the basics. On the other hand, if you truly understand the basics of science concepts, you're way ahead of someone who has memorized a college text from cover to cover, *without* understanding it. That said, I would like to mention one more thing about sound. I've given you a simple picture of what sound waves look like, because that makes the concepts easier to understand. Sounds in the real world, though, are complicated. Instead of looking like nice, smooth waves, a typical sound has a wave pattern that might look like Figure 7.18.

Figure 7.18

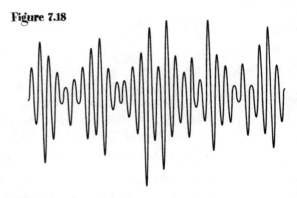

That might look scary, but there are ways to break down that scary looking mush of patterns into simpler sound waves.[12] Once you have things looking simpler, you can figure them out without too much trouble.

Chapter Summary

Record players and CD players transform mechanical vibrations into electrical signals, and then transform those electrical signals back into vibrations that produce sound. Telephones work in a similar manner.

The interaction of moving magnets and electrical currents makes the transformation from vibrations to electricity to vibrations possible.

Your ear utilizes all sorts of mechanical devices to transform incoming sound waves into electrical signals that travel to the brain via the auditory nerve. In doing so, a three-bone assembly called the ossicles acts like a lever and magnifies the effect of incoming sound waves.

Whenever you study parts of the human body, that's biology, and that means you run into a whole bunch of new vocabulary words, most of which are Latin. No, there won't be a test.

---

[12] The name for the process of reducing complex waveforms like this down into simpler patterns is known as *Fourier analysis*. I mentioned that in a previous chapter.

Applications

1. If you've ever heard your voice on tape, you probably were upset that your voice wasn't as low and resonant as you hear it when you talk. Bad news: the voice you hear on the tape is the voice other people hear. The reason you hear your own voice lower, and with better quality, is that you hear your own voice by conduction of the vibrations through the bones in your head, primarily your jawbone. That conduction enhances lower frequencies, making your voice sound lower to you.

2. Sometimes the ear doesn't work as it's supposed to, and you get hearing loss, or even deafness. There are two kinds of deafness. One is called *conductive deafness,* in which the outer and middle ears aren't working efficiently to get vibrations delivered to the inner ear. Lots of different things can cause conductive deafness, from damage to the eardrum to calcification of the bones in the ossicles. One way to correct conductive deafness is to use a hearing aid, which basically amplifies incoming sounds to magnify the vibrations that reach the inner ear. A hearing aid is just a tiny version of a microphone plus amplifier plus speaker.

    The second kind of deafness is called *nerve deafness,* in which the fibers or the tiny hair cells in the inner ear are damaged. If those two things don't work, it doesn't matter how much you amplify the vibrations that reach the cochlea, because those vibrations will never reach the auditory nerve. One solution to nerve deafness is a *cochlear implant.* A cochlear implant bypasses the damage and directly stimulates the auditory nerve. Here's how it works: The user wears a headpiece (actually, it's pretty small and usually fits behind the ear) that acts as a microphone, gathering in outside sounds. The headpiece converts the incoming sound to electrical signals (just like a phone) and transmits those signals to a speech processor. The speech processor amplifies the electrical signals, but it also adjusts the wave shapes so they can better stimulate the auditory nerve. I told you earlier that we don't fully understand how the brain interprets incoming electrical signals from the auditory nerve, but apparently different people respond better to some kinds of waveforms than others. There's an art to adjusting the speech processor so it maximizes the amount and quality of sounds the user can hear. The speech processor sends altered waveforms on to a receiver, which is basically a small magnetic disc (imagine that—a magnet!). The disc is surgically implanted just behind the user's ear, and it has an electrode attached that leads directly into the cochlea and directly stimulates the auditory nerve. I mentioned earlier how amazing the human body is. What is equally amazing is what people can create to help a malfunctioning body system operate!

3. You already know that the frequencies of sounds you can hear depend on the resonant frequencies of those thousands of tiny hairs in the basilar membrane, and also on the overall size and composition of the parts of your ear. It stands to reason, then, that animals with differently constructed ears can hear different ranges of frequencies. The simplest example of this is the fact that dogs can hear whistles that are too high pitched for humans to hear. Another example is the fact that my children can't seem to hear the particular frequencies I use when asking them to do chores.

Most references on the subject will say that the average human can hear sound frequencies as low as 20 Hz and as high as 20,000 Hz.[13] Those numbers vary from person to person, but most people don't fall too far outside that range. Knowledge of this average range of frequencies people can hear can keep you from getting hosed when you buy a sound system. If the salesperson wants you to pay extra for a system that produces sounds from 0.5 to 50,000 Hz, tell that salesperson to take a hike!

4. There is a scale we use to measure the loudness of sounds called the decibel scale. You've no doubt heard of this scale, so I'll try to give you some idea of how it works. For starters, let's talk about what leads to the loudness of a sound. You might expect that the amplitude (the size of the vibrations) of the sound wave determines loudness. That's almost correct, but not exactly. It turns out that the loudness (also known as the *intensity*) of the sound is related to the *square* of the amplitude of the wave. When you take a fairly ordinary range of numbers and square them, the results can end up being quite large or quite small. For the amplitude of the kinds of sounds we might hear every day, the intensities (which are related to the squares of the amplitudes) might range from 1,000,000 watts per meter squared to 0.00000000001 watts per meter squared.[14]

---

[13] The symbol Hz, used in honor of the physicist Heinrich Hertz, stands for a frequency of one cycle per second (in the case of sound, one wavelength per second).

---

[14] You'll just have to take my word for it that watts per meter squared are the proper units for intensity. Since we don't use those units anywhere else in the book, it just doesn't seem worth the time to go into a big ol' explanation at this point.

Glossary

Amplitude. The height of a wave that is actually half the distance from the lowest to the highest point on the wave. The amplitude of a sound wave determines the loudness of the sound.

Antinodes. Places on a standing wave that move the maximum amount.

Anvil. A part of the inner ear, also known as the incus, that helps transmit sound waves from the eardrum to the basilar membrane. Also a tool often used, ineffectively, by the Coyote against the Roadrunner.

Auditory canal. The part of your ear that leads from your outer ear (the pinna) to the eardrum.

Auditory nerve. A nerve that receives stimulations from tiny hairs and carries them to the brain, where they are interpreted as various sounds.

Basilar membrane. A tissue in the ear that runs along the center of the cochlea that contains thousands of fibers that react to incoming sound waves.

Beats. A "wah-wah" effect that's produced when two notes of similar frequency are played at the same time. There was an entire generation named after beats, but the name probably has very little to do with the "wah-wah" effect.

Chladni figure. The distinctive patterns you get when you vibrate a plate or other surface that is covered with sand or another similar material.

Cochlea. A long, coiled tube in the inner ear that contains all sorts of components to detect incoming sound waves.

Constructive interference. When two or more waves combine so as to create a result that is larger in amplitude than the individual waves.

Damping. The process of eliminating certain frequencies from a collection of sound waves.

Destructive interference. When two or more waves combine so as to create a result that is smaller in amplitude than the individual waves.

Doppler effect. A phenomenon in which a listener hears a higher pitch than that produced by an object when the object is moving towards the listener, and a lower pitch than that produced by the object when the object is moving away from the listener. This effect is named after a guy named Doppler. Go figure.

Eardrum. A membrane that vibrates when sound waves enter the ear. Ears don't work very well without eardrums. The technical name for the eardrum is the tympanic membrane.

Echo. The reflection of a sound. Bats make great use of echoes.

Eustachian tube. A tube that connects the inner ear with the sinuses and mouth. Its purpose is to equalize pressure between the inner and outer ear.

Fourier analysis. A mathematical procedure by which a complicated collection of waves can be broken down into individual waves of single frequencies.

Frequency. The number of wavelengths that pass a given point in a given time.

Fundamental. The lowest frequency and longest wavelength of a standing wave.

Hammer. One of a series of three bones in the inner ear that transmit sound from the eardrum to the cochlea. The Latin name for the hammer is the malleus. A commonly used household tool. Also the name of someone whose music career has seen better days.

Harmonics. The series of frequencies that are higher than the fundamental frequency in a set of standing waves.

Interference. The general name for what happens when two or more waves combine. Something cornerbacks hate being called for.

Linear density. The amount of mass per unit length of any object.

Longitudinal waves. Waves in which the direction of travel of the wave is in the same direction as the movement of the medium through which the wave is moving.

Medium. A substance through which sound waves can travel. Also the size between small and large.

Nodes. Parts of a standing wave that do not move at all. Also things that need blowing when people have codes.

Octave. An exact doubling or halving of the frequency of a note.

Organ of corti. Part of the basilar membrane that contains tiny little hair cells. These hair cells transmit the motion of fibers in the basilar membrane to the auditory nerve.

Ossicles. The collective name for the combination of the hammer, anvil, and stirrup.

Oval window. The place at which the stirrup connects to the basilar membrane.

Overtones. Another name for harmonics. This term is more commonly used in music.

Perilymph. The fluid that's inside the basilar membrane.

Pinna. The outer edge of the ear that collects sound waves.

Pitch. The perceived highness or lowness of a note. Except in rare cases, pitch of a note is determined by its frequency.

Resonance. The phenomenon in which one object responds strongly to outside vibrations.

Resonant frequencies. The frequencies to which an object responds strongly.

Round window. A membrane in the basilar membrane that allows fluid inside (the perilymph) to move freely.

Sonar. A method of using echoes to locate objects underwater.

Standing waves. A combination of waves and their reflections that make the pattern look as if it's standing still.

Stirrup. One of the bones in the inner ear that helps transmit sound from the eardrum to the basilar membrane. The Latin name for the stirrup is the stapes. Be sure to keep your feet in the stapes when riding a horse.

Tension. How tightly you hold a rope, string, or similar object. The kind of headache you can get trying to learn science.

Transverse waves. Waves in which the direction of travel of the wave is perpendicular to the direction of motion of the medium along which the waves are traveling.

Tympanic membrane. See "eardrum."

Wavelength. The distance in which a wave repeats itself, usually measured from crest to crest or trough to trough.

Index

Page numbers in **boldface** type refer to figures.

Stop Faking It!